Seeking Nontraditional Approaches to Collaborating and Partnering with Industry

Bruce Held
Kenneth P. Horn
Michael Hynes
Christopher Hanks
Paul Steinberg
Christopher Pernin
Jamison Jo Medby
Jeff Brown

Prepared for the United States Army

Arroyo Center
RAND

Approved for public release; distribution unlimited

The research described in this report was sponsored by the United States Army under Contract No. DASW01-96-C-0003.

Library of Congress Cataloging-in-Publication Data

Seeking nontraditional approaches to collaborating and partnering with industry / Bruce Held ... [et al.].
    p. cm.
    "MR-1401."
    Includes bibliographical references.
    ISBN 0-8330-3053-1
    1. United States—Military policy—Economic aspects. 2. Military-industrial complex—United States. 3. Defense industries—United States. 4. Civil-military relations—United States. I. Held, Bruce, 1958–

UA23 .S41634 2002
355.02'16—dc21

2001048218

RAND is a nonprofit institution that helps improve policy and decisionmaking through research and analysis. RAND® is a registered trademark. RAND's publications do not necessarily reflect the opinions or policies of its research sponsors.

© Copyright 2002 RAND

All rights reserved. No part of this book may be reproduced in any form by any electronic or mechanical means (including photocopying, recording, or information storage and retrieval) without permission in writing from RAND.

Published 2002 by RAND
1700 Main Street, P.O. Box 2138, Santa Monica, CA 90407-2138
1200 South Hayes Street, Arlington, VA 22202-5050
201 North Craig Street, Suite 102, Pittsburgh, PA 15213
RAND URL: http://www.rand.org/
To order RAND documents or to obtain additional information, contact Distribution Services: Telephone: (310) 451-7002;
Fax: (310) 451-6915; Email: order@rand.org

PREFACE

This report documents the findings of a study to help the Army understand how to better collaborate and partner with industry. It expands on a briefing, presented to the Assistant Secretary of the Army (Acquisition, Logistics, and Technology) (ASA(ALT)) in January 2000, describing nontraditional approaches for the Army to follow to collaborate and partner with industry using the concepts of public-private partnerships, venture capital funding, and government corporations.

The research was sponsored by the Deputy ASA(ALT) for Plans, Programs, and Policy and was conducted within RAND Arroyo Center's Force Development and Technology Program. The Arroyo Center is a federally funded research and development center sponsored by the United States Army.

The findings should be of interest to Army audiences concerned with collaborating and partnering with industry and interested in understanding the various means available to the Army to increase such collaborations and partnerships using nontraditional approaches.

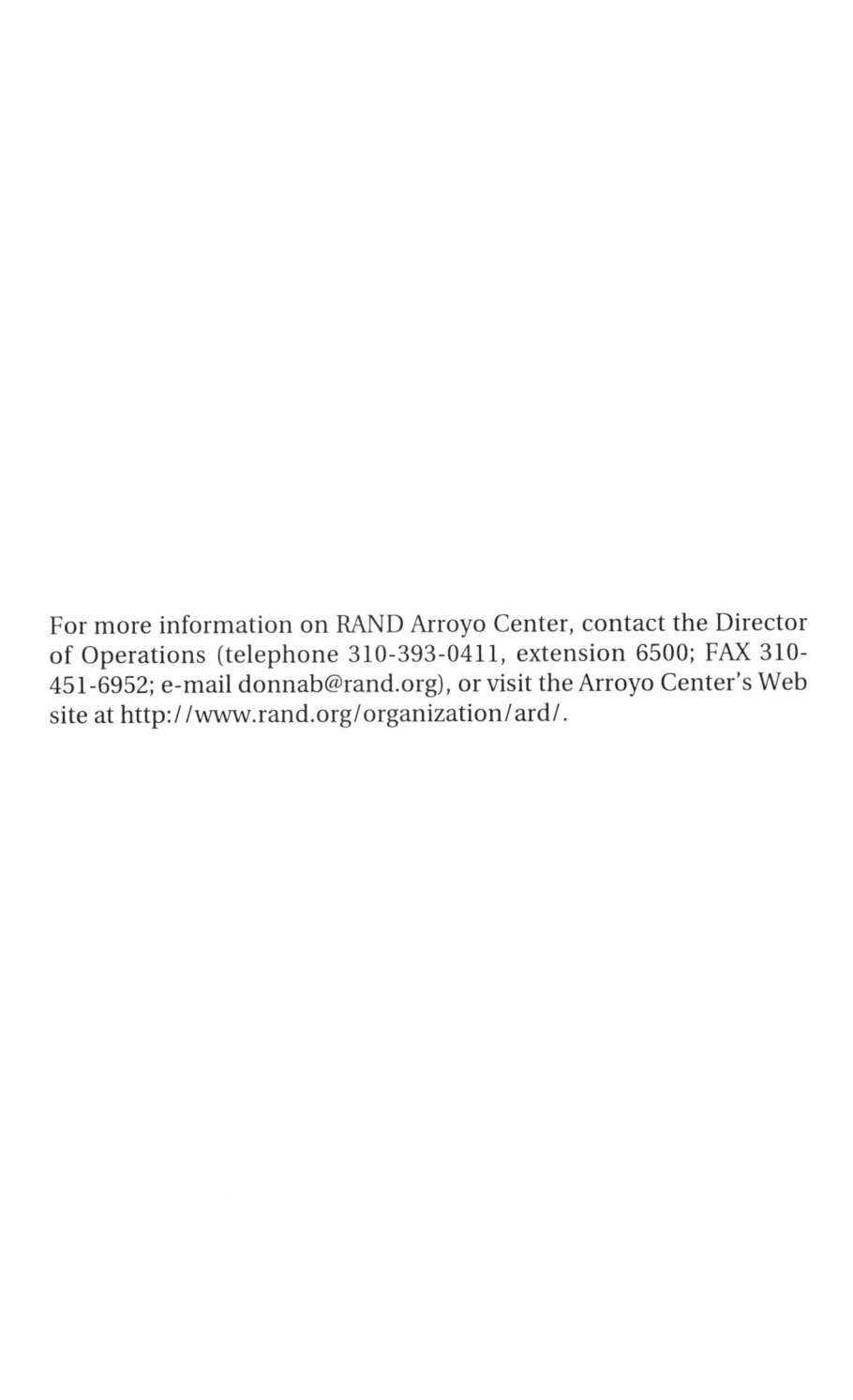

For more information on RAND Arroyo Center, contact the Director of Operations (telephone 310-393-0411, extension 6500; FAX 310-451-6952; e-mail donnab@rand.org), or visit the Arroyo Center's Web site at http://www.rand.org/organization/ard/.

CONTENTS

Preface	iii
Figures	ix
Tables	xi
Summary	xiii
Acknowledgments	xxi
Abbreviations	xxiii

Chapter One
  INTRODUCTION ... 1
  Background ... 1
    What Innovations Are Occurring in Industry? ... 2
    What Innovations Are Occurring in Government? ... 3
    What Innovations Seem Most Promising for the Army to Exploit? ... 4
  Objective ... 5

Chapter Two
  REAL ESTATE PUBLIC-PRIVATE PARTNERSHIPS ... 7
  Context ... 7
  What are PPPs? ... 9
  What Are the Benefits of PPPs? ... 11
    Improving the Army's Readiness Posture ... 11
    Leveraging Assets to Generate Revenue and Reduce Cost ... 12
    Leveraging Non-Army Resources for Improving and Adding to Army Assets ... 13
  Why Are PPPs Not Used More? ... 14

    The Army Is a Warfighter, Not a Realtor 14
    Competition with the Private Sector 15
    Bypassing the Political Process 17
    Security at Army Installations 18
  How Should PPPs Be Developed? 18
    Generating Appropriate Ideas for PPPs 18
    Valuing PPPs in Relation to Army Installations 20
    Choosing Mechanisms to Develop PPPs 27
    Leases 30
    Facilities-Use Contracting 31
    Special Legislation 31
  PPPs in Conclusion 32

Chapter Three
VENTURE CAPITAL APPROACHES 33
  Context 33
  Why Does the Army Have Difficulty Accessing the
    Commercial Technology Sector? 35
  What Has DoD Done to Address These Difficulties? 37
    New Tools Designed to Access the Commercial
      Technology Base 37
    Funding Specifically Aimed at Commercial
      Technology 40
  How Does a Venture Capital Fund Work? 40
  What Are the Benefits to the Army of Establishing a
    Venture Capital Fund? 43
    Can Exploit Innovation 43
    Can Be Used by Public and Large Private Organizations
      for Technology, Investment, and Nonfinancial
      Reasons 44
    Can Better Access Commercial Technology 45
    Can Leverage Non-Army Resources 46
    Can Provide a Return on Investment 46
    Can Give Rise to Entire Industries 47
  Possible Implementation Strategy for an Army Venture
    Capital Fund 48
    Establishing an Army Venture Capital Fund 48
    One Possible Model for an Army Venture Capital Fund . 50
    Appropriate Technologies for an Army Venture Capital
      Fund 52
    Integration with Other Army Technology Programs ... 52

Chapter Four
SPINNING OFF ARMY ACTIVITIES INTO
FEDERAL GOVERNMENT CORPORATIONS 55
Context.. 55
Federal Government Corporations 57
The Army and FGCs 64
FGC Areas of Concern 65
Potential Army Candidates for FGCs 66
  Army R&D Laboratories 66
  Army Depots 69

Chapter Five
CONCLUSIONS AND RECOMMENDATIONS 79

Appendix

A. DESCRIPTION OF LABORATORY MODELS
   CONSIDERED.................................. 81
B. DESCRIPTION OF DEPOT MODELS CONSIDERED 85

Bibliography 91

FIGURES

3.1. Venture Capital Funds Raised 47
4.1. Comparison of Organizations Along the
     Control Dimension 61
4.2. Comparison of Organizations Along the
     Cash Dimension 62
4.3. Comparison of Organizations Along the
     Customer Dimension 63

TABLES

2.1.	Value Indicators for Assessing Potential PPPs	23
2.2.	Value Indicators for Assessing Potential PPPs: Application to Picatinny	28
4.1.	Organization Laboratory Models Considered	68
4.2.	Organizational Depot Models Considered	72
4.3.	Depot Assessment Criteria	73

SUMMARY

INTRODUCTION

The Army has a growing need to collaborate and partner more with industry. When we look at the nexus between what is available to the Army in terms of the various innovations occurring in industry and elsewhere and what the Army can actually do to exploit those innovations, three innovative approaches emerge as promising candidates for collaborating and partnering with industry: (1) forming real-estate public-private partnerships (PPPs), (2) using Army venture capital mechanisms as a research and development funding and collaborating tool, and (3) spinning off Army activities into Federal Government Corporations (FGCs). While these three approaches seem promising, there is a need to understand more fully the potential and liabilities of each one, which is the goal of this document.

REAL ESTATE PUBLIC-PRIVATE PARTNERSHIPS

Although the Army has substantial real property holdings, it has diminishing resources to support or use those holdings. This mismatch forces the Army to make unpleasant choices. Using PPPs—characterized by a sharing of resources to achieve similar or, at least, not incompatible goals—can help the Army deal with the fact that it has valuable nonmonetary resources, including real property, that can help achieve Army requirements if used imaginatively and flexibly.

Previous RAND Arroyo Center research has shown that using PPPs can improve the Army's readiness posture, generate revenue, reduce cost, and leverage non-Army resources for improving and adding to Army assets. Despite these benefits, however, PPPs are not used much because they raise a number of concerns.

First, many argue that being so involved in PPPs directly impinges on the Army's core mission. However, the Army is in the business of real property maintenance whether it chooses to be or not, and maintaining its real property actually *detracts* from its core mission. Smart use of PPPs can reduce the Army's role in real property maintenance.

Second, many are concerned that making Army real property available for commercial use and development raises the potential for unfair competition with the commercial sector. Although the concern is real, it can be addressed by having the Army issue and enforce rules and guidelines for evaluating the "fair market value" of assets offered.

Third, many are concerned that in developing tools for implementing PPPs, attempts will be made to provide as much discretion as possible to the local level, while bypassing the political process. This concern can be addressed by carefully balancing the need for local discretion with requirements to address political concerns. While they are probably too strict, various safeguards are installed in current partnering tools to ensure that the political process is heavily involved in forming PPPs. The current safeguards can be used to help guide the development of more flexible partnering tools.

Finally, engaging in PPPs with Army assets implies that more people outside the military will have access to Army bases, thus raising security concerns. Although it is a legitimate concern, there are already mechanisms in place to deal with this issue. A related concern is that less desirable or even disruptive activities could find their way onto Army property; however, this can be addressed by establishing guidelines for permissible uses of Army real property.

Given that PPPs have numerous benefits and assuming that the legitimate concerns they raise can be adequately addressed, developing them entails pursuing three activities. First, appropriate ideas for PPPs must be generated. Instead of the current ad hoc approach, we recommend taking a proactive approach, starting with the premise that many good ideas for PPPs are being discussed at the installations and that, if tasked, the Army would end up with an impressive list of candidates.

Second, the Army needs to value PPPs effectively, which we argue should involve using a series of value indicators that include the quality of the local community, the mission of the installation, available capacity, criteria for success, and potential uses.

Third, the Army must decide among the tools available for implementing real property PPPs. Although they are limited, leases, facilities-use contracts, and special legislation show that these tools, used diligently and imaginatively, offer opportunities to create more PPPs with Army real property.

VENTURE CAPITAL APPROACHES

Over the past decade, the amount of resources the Army devotes to research and technology development has stagnated. Despite this, the Army's expectations and requirements for advanced technology continue to grow. Specifically, science and technology (S&T) capability is postulated as a central driver in the Army's planned transformation.

Given this asymmetry between resources and requirements, developers of Army materiel are forced to look to the commercial technology sector, which, unlike the Army (or, for that matter, the Department of Defense), has seen its research and development (R&D) spending quadruple in three decades and continues to grow at more than 4.5 percent per year. Unfortunately, like most DoD organizations, the Army has difficulty gaining and maintaining access to the advanced technology being developed in the commercial sector. The most prominent barriers to greater collaboration between the Army (and DoD) and industry are (1) intellectual property concerns, which combine with the fact that most companies do research for their own purposes, not as a service for hire, and (2) excessively bureaucratic requirements and the related distrust of government involvement and oversight in company affairs.

DoD has pursued some solutions to overcome these barriers, but they have not proved very effective. In particular, it has designed new tools, like Other Transactions (OTs) and Cooperative Agreements (CAs), to access the commercial technology base, but the Army has not made much use of them. The Small Business Innovative Research (SBIR) and the Small Business Technology Transfer (STTR) programs also address the problems to a certain extent and are relatively successful, but program limitations in terms of funding, timing, and nonmonetary resource availability have constrained their overall utility to the Army. Using a venture capital model for funding research and development of interest to the Army is one option for addressing the lack of access to the commercial technology sector. An Army venture capital fund that invests in companies and technologies that are of interest to the Army and have potential for commercial market penetration can provide significant benefits. In particular, an Army venture capital fund (1) can exploit innovation, (2) can better access commercial technology, (3) can leverage non-Army resources, and (4) can provide a return on investment (ROI).

Perhaps the best method for establishing an Army venture capital fund would begin in a small way with the Army partnering through an OT agreement with an established organization to begin work on a limited set of problems. The Army partner would organize and staff itself, if not already set up as such, to use venture capital as a tool for solving the problems in the partnership agreement. With an agreement in place and a small number of projects under way, the Army could then look for congressional endorsement and additional funding through the authorization and appropriations process.

Once established, an Army venture capital fund has to ensure that appropriate technologies are selected. First, the technology must have clear military and commercial applicability. Second, the Army must be a "power user" (i.e., have a requirement for a new product or technology ahead of other potential users). Finally, the technology must be "mature enough" to develop into a product or proprietary technology in the limited time and with the limited dollars that venture capital investing implies.

In addition, the fund must be integrated with other Army technology programs. One link is the need to find "sponsors" and users for the venture-backed technologies within the Army. Also, venture capital could be integrated into existing programs to make them better. For example, the SBIR program funds hundreds of technologies each year that are usually too immature for venture capitalists. An integrated venture capital approach could provide needed funding and support beyond that provided by the SBIR program. Likewise, the SBIR program could be a source of technologies for the venture capital fund, particularly if some of the SBIR awards are given with the fund's problem set in mind.

ARMY ACTIVITIES AND FGCs

FGCs were established over 200 years ago by Congress as a way to manage government-run operations that needed a high degree of autonomy and flexibility, more common to business-like activities, because these organizations were required to produce revenues to meet or approximate their expenditures. Since then, FGCs have become firmly established as an organizational model for governmental activities that have many attributes more common to a commercial business. Certain activities in the military (e.g., depots, research and development, transportation, etc.) may also be described as having commercial-like attributes and so may also be improved by adopting an FGC organizational structure.

Organizing certain military functions as FGCs may be attractive to the military because of the built-in flexibility. FGCs

- are free of many bureaucratic regulations such as the Federal Acquisition Regulation/Defense FAR Supplement (FAR/DFARS), the Civil Service regulations, the Competition in Contracting Act, various Office of Management and Budget (OMB) circulars, etc.;
- have flexibility of ownership in that they may be wholly or partially publicly owned with potential to be fully privatized;
- have flexibility of federal governance, i.e., freedom to assign board seats; and
- have flexibility in crafting language within their Articles of Incorporation that determine how the organization will be managed.

Given the inherent flexibilities in organization, personnel, and governance that FGCs enjoy, the military could take advantage of them and tailor FGCs to meet its specific needs.

At least three Army candidates for FGCs have been proposed: (1) Army chemical demilitarization, (2) Army R&D laboratories, and (3) Army depots. As part of the 1998 Army Materiel Command (AMC) Redesign Overarching Integrated Product Team (OIPT), the Army considered turning its chemical demilitarization operations into an FGC. Although the assessment was positive, no action was taken. At this stage in the demilitarization process, it may now be too late to consider making this organizational change. However, the other two FGC candidates are still timely and relevant, and the Army has not seriously studied them.

In both cases, RAND Arroyo Center assessed the value of different organizational models, including FGCs, in relation to a set of performance criteria. For the Army R&D laboratory analysis, which focused on AMC laboratories, the research looked at fifteen organizational models in terms of eight criteria using a Delphi evaluation approach with four rounds of ratings. The results indicated that the FGC model, as a candidate replacement for the current AMC laboratories, ranked in the highest of the four generic groupings, along with the federally funded research and development center (FFRDC) and the government-owned/contractor-operated (GOCO) models. By highest, we mean that the FGC, the FFRDC, and the GOCO models were the strongest models overall, with none of the eight performance criteria being challenging to achieve.

Analyzing Army depots, the research assessed six organizational models in terms of five generic categories of criteria using a traditional consensus-forming approach with the evaluators ranking the various alternatives after discussing them in an open forum. The results of the assessment indicate that the FGC, as a candidate approach dealing with issues in the current Army depot system, ranked in the highest overall among all the approaches.

There are three additional reasons why the FGC model is appealing for Army depot maintenance. The first reason is that using this concept removes the activity from the rigidity of the annual budgeting and appropriations process, when that rigidity conflicts with the basic nature of the business. The second reason for considering FGC approaches for the Army depot system has to do with the mandate facing the Army from the 1997 Quadrennial Defense Review (QDR) to eliminate 17,366 civilian positions by fiscal year 2004, 8,530 of which are supposed to come from AMC. By applying the FGC concept to its depot system, the Army could make reductions to its *government* civilian workforce without having to eliminate jobs. The third reason for looking at the FGC idea for the Army depot system is that senior Army leadership has already considered the concept and indicated its willingness to pursue it further—an important prerequisite for possible success.

CONCLUSIONS AND RECOMMENDATIONS

While the three collaborating/partnering concepts appear promising for possible Army exploitation, each one needs key issues to be resolved before the Army can seriously consider it for implementation. In the case of the PPPs, various implementation issues must be resolved within the Army, including whether the installations can propose financially sound concepts. In the case of the venture capital concept, its potential to meet the Army's technology needs must be addressed in further detail. Monitoring the status of the venture capital efforts undertaken recently by the CIA will help in this assessment. In the case of FGCs, the value of this model for Army laboratories and depots will depend on how much external commercial opportunity exists and further analysis on how to best structure continuing relationships with other Army organizations.

Once these key issues are satisfactorily addressed, the Army should create pilot programs to test the concepts. This approach is consistent with the new industry paradigm that argues that one learns more about something by acting on it (in this case, by establishing pilot programs) instead of, as in the past, waiting until it is thoroughly understood before acting.

ACKNOWLEDGMENTS

The authors wish to thank Bruce Waldschmidt, Director, Policy, ASA(ALT), for providing the study with valuable guidance, inputs, and insights; Brigadier General Stephen Gonczy, Defense Logistics Agency, for introducing us to the potential benefits of Federal Government Corporations; Richard Montgomery, a RAND consultant, for championing an Army venture capital fund option; George Kozmetsky, co-founder of Teledyne and emeritus head of the University of Texas' Innovation, Creativity and Capital (IC2) Institute, for sharing his insights on technology incubators and venture capital options; Todd Stevenson, ASA(ALT), for sharing his unpublished white paper on an Army leasing initiative pilot program; Elliot Axelband, a RAND consultant, for helping develop the Army venture capital concept; Carolyn Wong, a RAND staff member, for devising a methodology to assess the performance of different organizational models for the Army laboratories; and David Owen, a visiting exchange analyst from the United Kingdom's Defence Evaluation and Research Agency, for helping flesh out the candidate concepts.

Thanks are also proffered to the various innovators we talked with during the course of this study. They included real estate developers (James Harper, John Masterman, John Stainbeck), venture capitalists (Joel Balbien, Denny Ko, Gene Miller), business school professors (Jack Borsting, Paul Bracken), and advanced technology/technology transfer experts (Ken Dozier, Howell Yee).

Finally, we wish to thank Barbara Kenny for preparing the tables and figures and typing the manuscript.

ABBREVIATIONS

ACSIM	Assistant Chief of Staff for Installations Management
AIIC	Army Innovation Investment Corporation
AMC	U.S. Army Materiel Command
AMCOM	Aviation and Missile Command
AMTRAK	National Railroad Passenger Corporation
ANAD	Anniston Army Depot
AOR	Accumulated Operating Result
ARL	Army Research Laboratory
ARMS	Armament Retooling and Manufacturing Support
ASA(ALT)	Assistant Secretary of the Army (Acquisition, Logistics, and Technology)
AWCF	Army Working Capital Fund
BOD	Board of Directors
BRAC	Base Realignment and Closure
CA	Cooperative Agreement
CECOM	Communications-Electronics Command
CEO	Chief Executive Officer
CIA	Central Intelligence Agency
COE	Corps of Engineering

COSSI	Commercial Operations Support Savings Initiative
CPFF	Cost-Plus Fixed-Fee
CRADA	Cooperative Research and Development Agreement
CRAF	Civil Reserve Airfleet
DARPA	Defense Advanced Research Projects Agency
DFARS	Defense Federal Acquisition Regulation Supplement
DLR	Depot-Level Reparable
DMA	Depot Maintenance—Army
DMC	Defense Management Council
DoD	Department of Defense
DoE	Department of Energy
DUST	Dual Use Science and Technology program
DWCF	Defense Working Capital Fund
EDS	Electronic Data Systems
FAR	Federal Acquisition Regulation
FCS	Future Combat Systems
FDIC	Federal Deposit Insurance Corporation
FFRDC	Federally Funded Research and Development Center
FGC	Federal Government Corporation
FHA	Federal Housing Administration
FICO	Financial Assistance Corporation
FY	Fiscal Year
GAAP	Generally Accepted Accounting Principles
GAO	Government Accounting Office
GCCA	Government Corporation Control Act
GM	General Motors
GOCO	Government-Owned/Contractor-Operated

GS	General Service
GSE	Government Sponsored Enterprise
HQ	Headquarters
HQDA	Headquarters, Department of the Army
IOC	Industrial Operations Command
MHPI	Military Housing Privatization Initiative
MSC	Major Subordinate Command
MTDC	Massachusetts Technology Development Corporation
MWR	Morale, Welfare, Recreation
NOR	Net Operating Result
NTMS	Nontraditional Military Supplier
OIG	Office of the Inspector General
OIPT	Overarching Integrated Product Team
OMB	Office of Management and Budget
OT	Other Transaction
PEO	Program Executive Officer
PL	Private Laboratory
PLA	Patent Licensing Agreement
PM	Program Manager
PPBES	Planning, Programming, Budgeting, and Execution System
PPP	Public-Private Partnership
PVS	Prime Vendor Support
QDR	Quadrennial Defense Review
R&D	Research and Development

RCI	Residential Communities Initiative
RDEC	Research, Development, and Engineering Center
RDT&E	Research, Development, Test, and Engineering
RDX	Research Department Explosive (or cyclonite)
REFCORP	Resolution Trust Corporation
RFP	Request for Proposal
ROI	Return on Investment
S&E	Scientist and Engineer
S&T	Science and Technology
SBIR	Small Business Innovative Research
STTR	Small Business Technology Transfer
TACOM-ARDEC	U.S. Army Tank-automotive and Armaments Command—Armament Research Development and Engineering Center
TAS	Team Apache Systems
TECOM	Test and Evaluation Command
TRADOC	U.S. Army Training and Doctrine Command
TVA	Tennessee Valley Authority
U.S.C.	United States Code
USEC	United States Enrichment Corporation
USPS	United States Postal Service
WCF	Working Capital Fund

Chapter One
INTRODUCTION

BACKGROUND

In 1998, the Department of the Army asked RAND Arroyo Center to support an internal review of the U.S. Army Materiel Command (AMC). This review team, called the AMC Redesign Overarching Integrated Product Team (OIPT), was chartered to look at possible redesigns of AMC, and the Arroyo Center's role was to independently assess the technology-generation aspects of an AMC organizational redesign. A central issue facing AMC in the technology-generation area is how to keep its capability intact during a period of dramatic downsizing of its civilian acquisition workforce.

The Arroyo Center study found that in the science and technology (S&T) area, more cost-sharing and leveraging possibilities were possible.[1] The study introduced a framework for managing technology developments that depends on two dimensions: the technology's utility to the Army and its market breadth. When the Army's research and devleopment activities listed in the S&T budget were placed against the two management approaches that overlap commercial technology areas—"initiate" and "participate"—a number of potential opportunities to improve the Army's technology-generation capability through more collaborative efforts were identified.[2]

---

[1]Kenneth Horn, Elliot Axelband, Carolyn Wong, Ike Chang, Donna Kapinus, Paul Steinberg, "Redesign of AMC's Technology-Generation Function: Insights and Considerations," unpublished RAND research, November 1998.

[2]Carolyn Wong, *An Analysis of Collaborative Research Opportunities for the Army*, Santa Monica, CA: RAND, MR-675-A, 1998.

These findings were briefed to the Under Secretary of the Army, who recognized the importance of more collaborations and partnerships with industry and was interested in expanding beyond the traditional options the Army currently uses. Understanding what options exist means understanding the nexus between what is available to the Army in terms of the various innovations occurring in industry and elsewhere and what the Army can actually do to exploit those innovations given the relaxation of government control that has resulted from acquisition reform efforts.

What Innovations Are Occurring in Industry?

On the industry front, there are several ongoing trends. First, high-tech firms are focusing on innovations, not simply on research and development (R&D). The new paradigm is "searching not researching," which means that firms are seeking new ideas to exploit and are not necessarily performing the research themselves. This results in firms acquiring expertise from other firms as required, using technology licensing agreements, and forming various types of partnership agreements.

Second, many firms are putting their money into venture capital funds. Some have their own corporate in-house venture capital funds (such as Oracle or Siemens), while others prefer "arm's-length" funds (such as what Boeing has done in investing in venture capital opportunities).

Third, while a few firms with large conglomerate operations are forming new horizontal businesses to exploit the synergies of the conglomerate (e.g., General Electric), almost all firms are concentrating on core competencies and spinning off activities that are no longer consistent with corporate goals. A classic example of an industrial spin-off is General Motors (GM) selling off Electronic Data Systems (EDS) and part of its Delphi parts operation.

The innovative options that industry is pursuing fall into seven general categories: (1) spin-off, (2) strategic partnership, (3) venture capital, (4) merger/acquisition/consolidation, (5) consortium, (6) closure/liquidation, and (7) vendor consolidation (prime vendor). Six of the seven options have analogs in the military. Only venture capital does not have a direct analog, although, as we will discuss later, the CIA has recently taken a plunge into the venture capital arena.

What Innovations Are Occurring in Government?

On the government front, innovative initiatives have been implemented under the rubric of acquisition reform. As summarized in the "Defense Reform Initiative Overview," acquisition reform is intended to force the government to adapt better business processes, pursue commercial alternatives, consolidate redundant functions, and streamline organizations.[3] A primary function of the Army's acquisition reform is to "foster innovation [to create] creative and cost-effective solutions" by forming collaborative business arrangements, by relying on performance-based acquisition, by capturing and utilizing knowledge of the commercial marketplace, by enhancing competition, by consolidating requirements, and by incorporating innovative contractor incentives.[4] Thus, the scope of innovative approaches permitted under acquisition reform is large.

Probably the most innovative partnering concept in the military today as part of acquisition reform is the Military Housing Privatization Initiative (MHPI) and the Army's pilot program, the Residential Communities Initiative (RCI). The Department of Defense (DoD) estimates that about 200,000 military family housing units are old, lack modern amenities, and require renovation or replacement. Completing this work at the current funding levels and using traditional military construction methods would take 30 years and cost about $16 billion.[5] To improve military housing Congress enacted legislation at DoD's request authorizing a five-year pilot program to allow private-sector financing, ownership, operation, and maintenance of military housing. Under the MHPI/RCI, the Army can provide direct loans, loan guarantees, leasing and rental guarantees,

---

[3]Defense Reform Initiative 2000, "Defense Reform Initiative Overview" *(http://www.defenselink.mil/dodreform/overview/overview/htm)*.

[4]*United States Army Procurement Reform: 21 Century Vision—Business Advisors Contributing to Successful Achievement of Command Missions Through Innovative Business Agreements*, Army pamphlet published by the Army Acquisition Reform Directorate (SAAL-PR), undated.

[5]U.S. General Accounting Office, *Military Housing: Continued Concerns in Implementing the Privatization Initiative*, GAO/NSIAD-00-71, March 2000.

differential lease payments, conveyance or lease of existing property and facilities, and other incentives to encourage private developers to construct and operate housing either on or off military installations. In turn, the military service members use their housing allowance to pay rents and utilities to live in privatized housing.

The above example illustrates the extent to which the government is willing to change the way it does business to implement a cost-saving program. In addition, many positive changes have occurred as a result of acquisition reform. Among efforts to assist collaboration and partnering with industry, two notable changes stand out: the introduction of the "Other Transactions" (OT) authority with its flexibilities in constructing contractual agreements for R&D and prototyping, and the modification of the Federal Acquisition Regulations (FAR) to remove some of the (from industry's perspective) more inhibiting regulations.

What Innovations Seem Most Promising for the Army to Exploit?

Given the two sets of ongoing innovations, three approaches emerge as the most promising candidates for collaborating and partnering with industry: (1) forming real estate public-private partnerships, (2) using Army venture capital mechanisms, and (3) spinning off Army activities into Federal Government Corporations (FGCs).[6]

---

[6]The process used to sort through the various data and come up with these three approaches was thorough, although somewhat unstructured. It consisted of synthesizing individual analyses of relevant financial, legal, and political issues; discussions with leading "out-of-the-box" business thinkers representing four professional types (real estate developers, venture capitalists, business school scholars, and technology-transfer experts); and brainstorming sessions with study team members and consultants. This iterative process led to some interesting possibilities, as well as some false starts. As a result, it is difficult to reconstruct the exact course the analysis took.

In the end, however, three approaches emerged as promising candidates for collaborating and partnering with industry. Each appears to have significant potential payoff for the Army.

OBJECTIVE

While these three approaches seem promising on the surface, there is a need to more fully understand their potential and their liabilities. Chapters Two, Three, and Four explore each approach in more detail. Chapter Five draws some larger conclusions and insights and makes some recommendations. Appendixes A and B describe the models used in our analysis of spinning off Army laboratories and depots, respectively; the analyses themselves are described in detail in Chapter Four.

Chapter Two
REAL ESTATE PUBLIC-PRIVATE PARTNERSHIPS

Defense research-and-development spending is declining, fewer high-tech companies find it financially rewarding to help the military create weapons for the information age. Meanwhile, the Pentagon oversees a vast overcapacity of bases and other installations that consume billions of dollars, thanks to bureaucratic turf wars and congressional parochialism.

—Thomas Ricks
Wall Street Journal
November 15, 1999

CONTEXT

The U.S. Army controls roughly 12.7 million acres of land, making it one of the largest landholders in the country. While most of this acreage consists of open training and testing facilities, Army real property also includes more than 207,000 buildings, tens of thousands of miles of road, a million square yards of pavement, ports and runways, and utility structures—all of which require maintenance (source: Army Directorate of Public Works).

Unfortunately, although the Army is real estate rich, it is poor—in both funding and personnel—when it comes to the resources needed to manage and maintain its holdings. Since the end of the Cold War, resource allocation to the Army for real property maintenance has declined much faster than the rate at which the Army has been able to divest itself of its property. While the Army has transferred less than 2 percent of its land holdings over the last decade (Shambach,

1999),[1] its appropriations for real property maintenance have been halved.[2] The resulting shortfall is reflected in the difference between the estimated fiscal year 1999 (FY99) requirement for real property maintenance and the actual appropriation for that year: The Office of the Assistant Chief of Staff for Installations Management (ACSIM) estimated its requirement for real property maintenance at $2.26 billion but only received a $1.45 billion appropriation in FY99 (U.S. GAO, 1999b). In a similar though less dramatic manner, the Army's military construction budget has declined by about 25 percent since the end of the Cold War. As a result, many Army facilities now need substantial repair and maintenance. There are plans to increase resources for real property maintenance, but that increase is not guaranteed given the Army's multiple and simultaneous resource requirements.

Not only have there been budget cuts for real property maintenance, but the Army's personnel level also has dropped substantially in the decade since the Cold War ended: 50 percent in the case of civilian employees and over 30 percent in the military ranks (Office of the Under Secretary of Defense (Comptroller), 1999). Since the Army's real property divestments have been much smaller, we believe this implies that the Army's real property holdings are underutilized.[3] It also means that fewer personnel are looking after the same real property holdings.

The mismatch between real property holdings and the resources to support or use them forces the Army to make unpleasant choices. It can either spread and dilute limited maintenance resources over all requirements; prioritize its resources based on need; or identify those assets most important to its current mission and apply re-

---

[1] Since 1988, the Base Realignment and Closure (BRAC) process has identified 200,000 acres of Army land that will be transferred from the Army. Fewer than 50,000 acres have been transferred to date.

[2] In the late 1980s, real property maintenance averaged over $2.3 billion per year, while in the late 1990s it averaged $1.1 billion. These data are compiled from the Army Green Books, 1984–1989 and 1995–2000. The citation for the most recent Green Book is Assistant Secretary of the Army, Financial Management and Comptroller, *The Army Budget: FY01 President's Budget*, Washington, D.C.: Headquarters, Department of the Army, February 2000, p. 41.

[3] Some facilities are estimated to be utilized at less than one-third capacity (Shambach, 1999).

sources there, while ignoring or, at best, very significantly undermaintaining the rest. Each of these choices has significant negative impacts, including deteriorating infrastructure, loss of "surge" capability, declining morale, and significant future restoration cost. Limited budgetary relief will probably not solve the problem, and additional base closures are problematic in the near term.

In this chapter we examine public-private partnerships (PPPs), which can be part of a solution to the dilemma posed by underutilized and undermaintained facilities. We begin with a discussion of what PPPs are; then we explain their benefits and some of the reasons the Army has not used them more (along with some ideas for addressing these reasons). Finally, we look at a proposed approach for using real property PPPs.

WHAT ARE PPPs?

PPPs are arrangements in which the private and public sectors collaborate in some manner to achieve mutually beneficial goals. In general, PPPs differ from the traditional forms of public-private interactions in that traditional forms are characterized by a one-way flow of money—from the government to the private entity—with the private entity providing a service or product in return. Standard contracts to design weapon systems would be an example of traditional public-private interaction. In such interactions, the private party's incentive is normally to maximize profits realized from the transactions, while the government's goal is to obtain specific services and products for the money expended.

Unlike these traditional public-private interactions, PPPs are much more flexible. They are characterized by a sharing of resources to achieve similar or, at least, not-incompatible goals. For the Army, the most important aspect of PPPs is that they can deal with the Army's nonmonetary resources, including real property, that are valuable and can be used to achieve Army requirements if used imaginatively and flexibly.

Real property PPPs are not new to the federal government. For example, the General Accounting Office (GAO) studied six real estate partnerships entered into by the U.S. Postal Service and the Department of Veterans Affairs (U.S. GAO, 1999c). These partnerships operate on several different models, but all involve bringing in private real estate developers to develop, manage, and operate publicly owned property, and they all provide tangible benefits to both partners. The government receives income from its underutilized property, as well as improved maintenance of that property. The private partner earns profits and gains access to attractive business locations.

The Civil Reserve Airfleet (CRAF) program is a current example of a PPP in the Department of Defense. CRAF is a public-private partnership in which commercial airline and air freight companies set aside specified aircraft for federal use during national emergencies. Occasionally, these aircraft are even modified to accommodate the types of materiel and missions required during call-up. As compensation, the involved companies are guaranteed a portion of the government's peacetime business.[4] CRAF is not a case involving private-sector use of government assets, but its structure is similar to the real estate partnerships we recommend in a fundamental way: resources are used and maintained in peace by the private sector and are thus quickly available in times of national emergency ("Civil Reserve Air Fleet,"1999).

A more direct example of how public-private real estate partnerships enhance readiness is the Navy's lease of part of its Port Hueneme facility to the Mazda Corporation. Mazda leases underutilized facilities from the Navy and, in return, helps to maintain the facility. This arrangement helps the Navy ensure that the facility will be ready immediately should expanded capability be necessary to address a national emergency.[5]

---

[4]CRAF participants are allocated a portion of the government's passenger and air freight business based on the total passenger and lift capability of the aircraft they enroll in the CRAF program. These aircraft are contractually promised to the government for use during call-up. As a result, the government avoids the cost of acquiring, maintaining, and manning the aircraft during peacetime. The government also receives reasonable and stable prices for air services during peacetime, though these government-negotiated fares may be higher than commercially available, advance-purchase, nonrefundable fares.

[5]Private communication with Contracting Officer, Port Hueneme Navy CBC, on Mazda Lease, Contract Number N62474-96-RPOQ04, 1997.

WHAT ARE THE BENEFITS OF PPPs?

These successes indicate that the Army could also substantially benefit by entering into similar arrangements. Previous RAND research (Chang et al., 1999) has identified some of these benefits: (1) improving the Army's readiness posture, (2) leveraging assets to generate revenue and reduce cost, and (3) leveraging non-Army resources for improving and adding to Army assets. We discuss these three items below.

Improving the Army's Readiness Posture

Of all the benefits of public-private real estate partnerships, the most important may be that using them may improve the Army's readiness posture. First, since real property maintenance is not an Army core competency, involving a larger percentage of the Army's personnel in this task reduces their contribution to the Army's primary mission of preparing for, deterring, and fighting the nation's wars. Any measure, such as a real estate public-private partnership, that improves the Army resource allocation toward its core missions should improve its readiness posture.

Second, as mentioned earlier, the Army's civilian and military personnel strength has declined dramatically over the last decade, and its budget for real property maintenance has also decreased steeply. Since its real property holdings have not declined at the same rate, this implies that either a greater percentage of the Army's workforce is involved in maintaining property or the property is not being maintained effectively. Letting underutilized property deteriorate will eventually detract from the Army's readiness posture. In many cases, property underutilized during peacetime is property that may become critical during times of national crisis. During a general mobilization, for example, active Army installations could be required to rapidly expand to accommodate the influx of National Guard and Army Reserve soldiers who come to train and to prepare for deployment. Likewise, facilities used for Army acquisition purposes may need to expand when research, development, and procurement activities are accelerated. If the Army must expand into facilities that require renovation and rebuilding, that expansion will be delayed or inhibited. Allowing peacetime use of currently underutilized facilities by private entities could help ensure their readiness for the Army in times of need.

Allowing underutilized facilities to deteriorate affects readiness in another way. When maintenance is disregarded, the cost associated with owning deteriorating property is not avoided—it is merely deferred. The facilities will need repair or removal at some point in the future, and potential environmental and safety hazards associated with deteriorating facilities could increase future rehabilitation costs. In these cases, the costs associated with neglected facilities will affect both future Army budgets and future readiness.

Leveraging Assets to Generate Revenue and Reduce Cost

Public-private real estate partnerships can also be an important method for leveraging the Army's property portfolio to improve revenue or reduce maintenance or other costs. As mentioned earlier, these partnering arrangements are not new at the federal level. For example, the U.S. Postal Service entered public-private real estate partnerships that developed valuable property it owned in New York City and San Francisco. In each case, the private partner developed the property and enhanced the facilities used by the Postal Service. The private partners also substantially expanded the facilities, found tenants, and now pay the Postal Service a portion of the collected rents (U.S. GAO, 1999c). The benefit to the Postal Service in both of these cases has been twofold: facilities have been renovated, and an income stream has been generated from what was once underutilized property.

The Armament Retooling and Manufacturing Support (ARMS) program is an example of an Army partnering program already producing tangible results in terms of lower facility maintenance costs and revenue generation. This program allows facility contractors to lease dormant facilities at Army ammunition plants to commercial enterprises. One recent evaluation of the program notes that "with the public sector investment in ARMS to date totaling $170 million, the Army has recovered $125 million and has resulted in over $2.1 billion in economic impact" (Open Enterprise, 1999).

Leveraging Non-Army Resources for Improving and Adding to Army Assets

Public-private real estate partnerships also provide a way for the Army to leverage resources otherwise unavailable to it. For example, the Residential Communities Initiative (RCI), the Army's version of the Military Housing Privatization Initiative (MHPI), is a pilot program that allows contracts with private developers for constructing new housing and renovating existing quarters on four installations. In a traditional construction contract, the services of the developers would be paid for and the Army would assume ownership and maintenance of the new and renovated housing units. As a PPP, however, the developers provide much of the RCI's funding. In return, they are guaranteed income from the projects for a number of years and may even gain ownership of the property in some cases. Additionally, the Army can provide loan guarantees, can provide loans at advantageous interest rates, or can even invest in the development companies.[6] Although the Army is just beginning to use this legislation, the benefits of the program can be inferred from recent successes in similar development projects undertaken at Corpus Christi Naval Air Station and Country Manor in Everett, Washington,[7] which are hoped to provide adequate housing for the increasing number of married soldiers with families.

Another obvious candidate for public-private partnerships is the Army's industrial facilities. For instance, a private partner could install state-of-the-art production facilities in an effort to improve efficiency or could upgrade existing Army equipment. Similarly, Army laboratory facilities could be improved through partnerships with high-technology firms. A recent study found that 35 percent of the laboratories and 52 percent of the test and evaluation centers in the DoD are excess (U.S. GAO, 1998b). These facilities are used for research in a wide range of disciplines, including areas with obvious private analogs, such as electronics and aerospace. If the Army can find private partners willing to share the cost of developing and redeveloping these research facilities, the problem of excess property

---

[6]Residential Communities Initiative homepage, *http://www.rci.army.mil/ program.*

[7]For details on these projects, see the Military Housing Privatization Initiative Web site, *http://www.acq.osd.mil/installation/hrso/.*

at these installations could be reduced, the Army would gain access to new research capabilities, and the value of its real estate holdings would be increased.

WHY ARE PPPs NOT USED MORE?

Despite these benefits, real property PPPs are not used more within the federal government generally and by the Army specifically. This indicates that there are issues and reservations associated with their use. We identify four such issues below and address the concerns.

The Army Is a Warfighter, Not a Realtor

A principal objection to the Army's use of PPPs is the feeling that the Army is not in the "business of business": its mission is to fight and win the nation's wars, and that does not include managing private tenants and real property portfolios to create revenue. The concern has two parts; the first is that involvement in PPPs directly impinges on the Army's core mission, while the second is that such involvement may support activities that have nothing to do with the core mission. We address each concern in turn.

As for the first concern, one of the main reasons for entering into PPPs in the first place is to allow the Army to *concentrate* on its core mission of preparing for and conducting combat operations. While the Army may not be in the "business of business," it is in the business of real property maintenance whether it wants to be or not, but smart use of PPPs can reduce the Army's role in real property maintenance. Using private developers whose core competency is in real property maintenance, development, and management can minimize the Army's role in these functions and allow it to apply its energy and resources to its primary missions.

As for the second concern, issues may also arise from PPPs that do not on the surface look to support the mission of the Army. For instance, some of the contractors under facility-use contracts in the ARMS program are producing goods that do not directly benefit the Army mission (e.g., consumer products). However, the Army reaps substantial benefits from the fact that the commercial entities are operating on its installations. Such activities help maintain infrastructure, enhance the local economic base that would support the installation in an emergency, and maintain a local workforce. Additionally, resources obtained through PPPs, in the form of either cash or in-kind services, are certainly of benefit to the Army.

Since some of the activities in real property PPPs may not have direct military utility, there is concern that less-desirable or disruptive activities could find their way onto Army property. Although such a concern is real, it can be addressed. For example, the Norfolk Willoughby land development case addressed it by listing businesses not allowed to lease Navy land. That list included such businesses as offtrack betting and adult bookstores and was circulated with the Request for Proposal (RFP).[8] Similar guidelines can be developed, either directly on a case-by-case basis or through a "zoning" committee which ensures that only businesses consistent with Army functions be allowed on its installations.

Competition with the Private Sector

Another concern is that making Army real property available for commercial use and development raises the potential for unfair competition with the commercial sector. This concern is both practical and philosophical. We address each concern below, starting with the practical one. From a businessman's standpoint, the Army enjoys certain advantages. First, assets in Army use are provided by the government and are not typically valued by the Army as a commercial entity would value them. Additionally, since Army property is federal property, it is not subject to the many local and state rules that apply to competing properties. Local zoning regulations are an example of a regulatory restriction that does not apply to federal property. If such advantages are exploited, the end result can be an unfair advantage in the competitive market.

While the concern about unfair competition is real, it may be addressed in several ways. In some cases, the PPP concerns a business for which there is no commercial competition. This is the case at the Military Ocean Terminal in Sunny Point, North Carolina (ArmyLINK

---

[8]Authority for the development of the Willoughby site fell under Public Law 102-190, section 2838.

News, 2000). There, an agreement between the commercial explosives industry and the Army allows for the commercial shipment of explosives through the port. In return, the Army earns revenue, gets upkeep for the facility, and gives its terminal operators experience in doing the hazardous work. Competition is not a big issue because of the extreme limitations put on the shipment of explosives through commercial U.S. ports. The alternative to shipping through Sunny Point is for the commercial firms to ship explosives to Canada and then use ground transportation to get them to the United States. However, the lack of alternative unloading points in the United States makes a virtual nonissue of unfair competition in this case.

Other PPPs will concern businesses that are doing business with the Army or federal government. Defense contractors may wish to locate office space, laboratories, and production facilities on Army property to be nearer to their customers. This already occurs to a limited extent, and PPPs could greatly expand this practice. While competing non-Army facilities may be available nearby, the use of government property by businesses working for the government can reduce product and service costs to the government. This makes a strong case for these kinds of PPPs.

Most of the potential PPPs are cases where commercial alternatives exist and the potential user of Army real property does little business with the Army. For these types of potential PPP, rules and guidelines for evaluating the "fair market value" of assets offered by the Army must be issued and enforced. But these guidelines must be flexible enough to take into account changing market conditions (for example, the effects of zoning or the sometimes rapid changes in real property values that accompany local economic changes) and alternate property uses.[9] Maintaining a requirement and mechanism for conducting fair-market-value assessments should temper concerns that the Army is using its property noncompetitively.

---

[9]10 U.S.C. 2667 already requires fair-market assessments. Currently, when the Army wishes to lease real property, the Army Corps of Engineers conducts an assessment to determine the property's fair market value. The regulatory framework for conducting these assessments may need to be reexamined. For example, current regulations (32 C.F.R. 644.41) require that the assessed value be based on "highest and best use." This requirement may be somewhat inflexible and could result in bureaucratic rejection of otherwise worthwhile projects.

Beyond the practical concern discussed above, there is also a philosophical concern about the government even being involved in the private sector in this capacity. Clearly, an argument can be made that the government has no role, other than a regulatory one, in the commercial marketplace. The source of this belief is based to some extent on the unfair advantage issues raised above. However, perhaps a more important fear is that government entry into commercial enterprises creates conditions that promote corruption. Constructing a transparent and open process for developing PPPs is therefore essential. Involving the political process can provide governance for PPPs, but exposing the agreements to the public will help to both disseminate the public good of the partnerships and ensure an adequate public approval and sufficient public scrutiny.

Bypassing the Political Process

For PPPs to have much of an impact, there must be significant use of public assets by the private sector. The more extensive this use, the more PPPs move into the political realm. Contradictorily, the more discretion that is allowed at the local level, the likelier it is that PPPs will be successful. This naturally raises a concern that in developing tools for implementing PPPs, attempts will be made to provide as much discretion as possible to the local level, while bypassing the political process. This concern can be addressed only by carefully balancing the need for local discretion with requirements to address political concerns. Various safeguards are installed in current partnering tools to ensure that the political process is heavily involved in forming PPPs. For example, there are already congressional notification requirements for 10 U.S.C. 2667 leases. These include an annual report to Congress detailing all new leases and changes to existing leases, as well as notification of intention to lease in certain cases or to spend lease receipts that are valued above a minimum threshold. Providing detailed guidelines with flexible partnering tools is also necessary, but carefully balancing local discretion with congressional oversight will continue to be the best method of addressing any political process concerns.

Security at Army Installations

Engaging in PPPs with Army assets implies that more people outside the military will have access to Army bases, thus raising security concerns. While this is a legitimate concern, there are already mechanisms in place that deal with this issue. Many, if not most, Army posts are already open to the public to provide access to facilities such as museums, clubs, or golf courses. Areas that require security are simply closed to the public. Other measures that are currently employed for security, such as registration of vehicles, could be expanded to the additional workforce. Perhaps the best way to address this concern is to understand that security must be a part of any installation's real property PPP plans.[10]

HOW SHOULD PPPs BE DEVELOPED?

Given that PPPs have numerous benefits and assuming that the legitimate concerns they raise can be adequately addressed, the next issue has to do with how to develop them within the Army. This entails three activities: (1) generating appropriate ideas for PPPs, (2) valuing them effectively, and (3) deciding on the mechanism to actually develop them. We discuss each of these activities below.

Generating Appropriate Ideas for PPPs

Until now, the Army has used an ad hoc approach for finding public-private real estate partnering opportunities. By ad hoc, we mean a generally passive approach in that the Army staff neither encourages partnership ideas nor provides resources to develop ideas that may evolve. Instead, entrepreneurial officials at installations, sometimes with local community support, must conceive of and develop ideas independently and forward these up a generally unenthusiastic chain of command. Development of these ideas takes time and resources that are hard to come by at the local level. Additionally, and as we shall discuss further on, restrictions with the existing outlease statute impede business arrangements with potential tenants. When

---

[10]In another vein, the fact that military installations are so secure may even be made into an asset for PPP purposes by attracting commercial firms that value the added security associated with military bases.

the low level of Army staff enthusiasm is combined with a lack of resources for real property business development and a legal structure biased against partnering, it is not surprising that few PPP real estate ideas have surfaced or been completed.

To date, only Fort Sam Houston's and Picatinny Arsenal's partnership plans have progressed very far. Fort Sam Houston's plan to develop several large buildings with a private partner has moved to the point where a developer has been selected and work may begin soon, though this progress has been several years in the making. Plans for leasing and developing three small- to medium-sized buildings at Picatinny Arsenal, New Jersey, are also under way. Congress has been notified and a solicitation has been drafted, although it has not yet been issued. Other projects in the queue include the construction of a contractor support facility at Fort Leonard Wood, Missouri, the development of a hot-weather test track at Yuma Proving Ground, Arizona, and the lease of manufacturing equipment and facilities at Rock Island Arsenal, Illinois. Developing an airfield at Fort Hood is also funded and under way, but this project is not a PPP in the sense we mean, since funding for the project is public, though not Army, money.[11]

The primary reason cited for the current "go-slow," ad hoc approach to Army public-private real estate partnerships is that congressional support is tepid and cautious. To overcome congressional resistance, Army Staff personnel believe it is very important to ensure that every proposed project be unambiguously legal and have consensus agreement within the Army. By this rationale, the task of testing the legality of various concepts, creating the business case, and developing support within the Army has taken a great deal of time and effort that should not be squandered by advancing more projects than could be adequately handled before the success of these partnerships has been shown.

While there is some merit to this argument, using the ad hoc approach described above risked missing the best ideas and potentially supporting others that are marginal from a business standpoint. Thus, we recommend a more proactive approach. It starts with the

---

[11]The Texas Department of Transportation and the Federal Aviation Administration are both providing funding for this project.

premise that there are many good ideas for PPPs being discussed at the installations and that, if tasked, the Army would end up with an impressive list of candidates. This list could then be winnowed down to several of the best ideas, which could be the basis for expanding the Army's use of public-private real estate partnerships.

The critics of this solicitation approach argue that the installations are not well versed in financial matters and would not be able to identify a good idea. We believe that the installations are either savvy enough to be aware of their potential winners or could develop the business sense required to make wise business decisions concerning public-private real property partnerships. This is particularly so if commanders have the tools, such as valuing guidelines (discussed in the next subsection), and the resources to follow up and determine the efficacy of various ideas.

Additionally, if the Army were to adopt a more proactive approach to public-private real property partnerships, we would anticipate a more formal approach at the Army Staff level that would provide additional guidance and training for installation commanders and their staffs. The guidelines and training would focus on providing direction about the types of businesses that would be compatible with Army real property, on developing innovative real property partnership tools, on developing installation property management offices, and on helping installations build sound business approaches.

Providing this level of guidance and assistance requires resources, but the payoff seems worthwhile. By using the proactive solicitation approach, the winning concepts would be pushed forward, and because the probability of getting good ideas increases, it becomes easier to obtain congressional approval. As more winners are implemented, the process should gain momentum and support, both within the Army and with Congress.

Valuing PPPs in Relation to Army Installations

For PPPs to be most successful, installation commanders should work with real estate developers. Since the bases must continue to support their military missions, commanders and other installation officials must work with private real estate developers to help them appreciate the unique missions and requirements of Army installations. However, the Army personnel must also be able to recognize their installation's potential private market value in light of the unique missions and requirements the military places on the property. This requires Army personnel responsible for real property PPPs to be able to understand the value of their installation from the perspective of potential private partners who are considering investing in it.

To help establish an appropriate mindset for evaluating Army property, there are numerous sources of data and information that should be consulted when writing proposals and negotiating with private developers. For example, local chambers of commerce, local and state governments, real estate organizations, and various industry associations typically maintain the sort of information that can greatly help in developing value indicators, business plans, and development schemes.

In this subsection, we discuss value indicators for evaluating real property PPPs. We then illustrate the concept with a short case example of Picatinny Arsenal.

Value indicators for real property PPPs. Real property partnerships between the Army and private developers depend heavily on the match between the local community's needs and the assets available on the installation. Fortunately, the history of public-private agreements in other government agencies provides several guidelines for the Army to follow when considering whether private development may be appropriate for a particular location. Based on a number of case studies, existing RFPs, and other sources, the following value indicators should be considered when evaluating real property partnerships.[12]

- **The quality of the local community.** A thorough evaluation of local economic and demographic conditions will provide impor-

---

[12]Case studies include National Park Service agreements for the development/renovation of historic buildings, Department of Veterans Affairs ventures for leasing land for office construction, and U.S. Postal Service leasing of commercial office space to private developers in desirable urban areas. Detailed case studies from each of these agencies are included in U.S. GAO (1999c). Army-specific ventures such as the ARMS and RCI programs are also instructive.

tant details about growth prospects, emerging industries, and likely responses from the private sector.

- **Mission of the installation.** The installation's mission must be considered to determine how and what kind of real property partnerships will fit within that mission.
- **Available capacity.** Available capacity should be identified by a real property assessment at every installation. That which is least critical to the Army's mission should be targeted for real property PPPs.
- **Criteria for success.** Determining Army goals when deciding whether to pursue real property PPPs is critical. Goals are an essential part of planning, negotiating, evaluating, and managing PPPs. Nonexistent or vague criteria for success will only hamper the development of PPPs.
- **Potential uses.** Based on these first four stages, the most promising matches between local private-sector demands, available Army facilities, and Army goals can be identified and pursued.

Table 2.1 summarizes these value indicators and provides some more specific detail about relevant questions and relevant variables.

A case study of Picatinny Arsenal. We use Picatinny Arsenal, located in northern New Jersey, to illustrate the potential for real property PPPs.[13] As a case study, Picatinny Arsenal has two advantages. First, personnel there have aggressively pursued both R&D PPPs—including Cooperative Research and Development Agreements (CRADAs) and Patent Licensing Agreements (PLAs)—within an innovative technology transfer center, and, more recently, real property PPPs. Arsenal officials are soliciting developers and taking the bureaucratically required steps to partner with the private sector

---

[13]The following list of organizations is an example composite of the types of entities that should be consulted when considering an installation's potential for a successful PPP. These organizations provided useful information on the Picatinny Arsenal case study: Morris County Chamber of Commerce; New Jersey Economic Development Authority; Picatinny Technology Transfer Programs; Picatinny Technology Innovation Center; and New Jersey Chapter of the National Association of Industrial and Office Properties.

Table 2.1
Value Indicators for Assessing Potential PPPs

Indicator	Relevant Question(s)	Relevant Variables
Quality of local community	What are the relevant economic and demographic conditions in the community?	Projected economic growth Projected employment growth Projected growth industries Projected population growth Projected labor force growth Average education Office/industrial vacancy rates Price/square foot of office space Distance to metropolitan area Major local university Major local employers Special development incentives
Quality of installation	What are the installation's major land, construction, and infrastructure assets?	Buildings Type of service conducted Equipment Available workforce Land Utilities
Available capacity	How can current installation assets be consolidated, relocated, or restructured to accommodate a PPP?	Buildings
	What military assets can be used for the PPP?	Dual-use equipment
Criteria for success	Is the PPP in consonance with the installation's military goals?	Army mission at installation
	Are there other benefits to the PPP?	
	Does the PPP meet the criteria for success as outlined by the February 1999 GAO report?	Catalyst for change Statutory basis exists Detailed business plan Coherently organized structure Stakeholder support
Potential uses for installation assets	What types of private enterprise should be encouraged to engage in a PPP?	Qualities of potential businesses should be consistent with installation assets capabilities

about some real property on the arsenal. The proposed actions are somewhat limited and, thus, provide some perspective concerning the Army's current approach to PPPs.

The second reason for using Picatinny as a case study is that intuition indicates that it ought to be a prime candidate for more substantial real property PPP development. It is located in a relatively densely populated area where the economy is strong. It is near a large city, and, because of workforce reductions over the last decade, it would seem to have assets available for development. Testing our intuition is a matter of applying the value indicators developed above and detailed in Table 2.1.

Quality of local community. An initial assessment of the local community's demographics and economic situation supports the idea that Picatinny Arsenal is a good candidate for the development of real property PPPs.

Picatinny Arsenal is located in Morris County, New Jersey, 28 miles west of New York City and 60 miles north of Philadelphia. Morris County is one of the region's least densely populated areas, but it has become increasingly suburban in recent years. Existing development consists of corporate headquarters, business parks, R&D laboratories, and even agricultural operations in the western part of the county. The county's highly educated labor force, efficient transportation infrastructure, and high quality of life reflect the primary benefits for attracting businesses. The highly skilled population of the county should draw high-paying, white-collar jobs from companies seeking to relocate from major metropolitan areas such as Newark or New York City, while the availability of vacant land and relatively low cost of living should allow existing local business enough room for expansion. Morris County has also been designated as one of New Jersey's three Foreign Trade Zones, which allows companies to defer duty payments on imported goods. Finally, the county offers generous tax credits for attracting new development and regularly provides detailed economic and demographic projections in *Morris County's Electronic Factbook.*

Quality of installation. In addition to being well located, Picatinny Arsenal itself appears attractive from a development standpoint. The arsenal contains a diversity of assets that complements its primary mission of armament research and engineering. There are over 1,000 buildings and structures at Picatinny, which includes numerous laboratories, office space, storage facilities, a museum, a conference center, two restaurants, and residential housing. Its extensive transportation and utility infrastructure exceeds the quality of many business parks. Over 3,000 employees work on a range of projects, including ballistics, advanced propulsion, and environmental sciences. Picatinny also contains a technology-transfer center for licensing Army technology to private companies for dual-use projects, with over 20 tenants currently engaged in CRADAs. The Picatinny Technology Innovation Center serves as one of six business incubators in New Jersey and provides training, administrative, and information technology support to start-up companies. In general, Picatinny offers a comprehensive support network for virtually any business seeking to expand its current operations or simply to relocate to more desirable surroundings.

Available capacity. The buildings on Picatinny's land can benefit both private and Army interests.

The availability of some buildings for outlease is indicative that recent workforce reductions have resulted in substantial excess interior and exterior space at Picatinny. Making the best use of this available capacity, both as it serves arsenal functions and in terms of its development potential for real property PPPs, would be the goal of a commander who proactively pursues real property PPPs. The process to achieve this goal would start with an evaluation, both by type and by quantity, of the essential real property requirements for Picatinny's mission. Based on this evaluation, consolidating and optimizing Army use of the arsenal, while making contiguous blocks of property available for commercial development, should enhance Army goals and improve the arsenal's development potential. For instance, government operations could be concentrated in a "downtown" area, opening up other developed parts of Picatinny Arsenal to real property PPPs.[14]

Criteria for success. Criteria should consider and measure the economic, operational, and community benefits that the installation, its

---

[14]In contrast, the current proposal offers to lease several relatively small, separately located buildings.

surrounding area, and the private partner can gain through a well-designed partnership. Criteria should be quantitative to the extent possible, but it is likely that many important indicators of success are, at best, measured subjectively. It is also important to note any "show-stoppers": categories of partnerships that will not be allowed, regardless of criteria such as economic benefit. The success criteria below are pulled from the Picatinny Arsenal's current Request of Application of Leasing. Suggested metrics are included.

- Utilize Picatinny assets consistent with the requirements and mission of U.S. Army Tank-automotive and Armaments Command—Armament Research Development and Engineering Center (TACOM-ARDEC) and its collocated customers.
  Metric: Occupancy rates and subjective evaluation of tenant relevance to Picatinny/Army requirements and missions.

- Provide state-of-the-art facilities for the Army.
  Metric: Value of new construction and installed equipment.

- Provide good stewardship of the real assets located on the installation and, to the extent possible, defray installation operating costs.
  Metric: Value of maintenance provided by tenants, real property maintenance back orders, and resources available to Army-maintained property.

- Provide Picatinny with access to technology that will assist in mission accomplishment.
  Metric: Number of CRADAs, PLAs, and R&D contracts with tenants.

- Maintain positive relations with the communities surrounding the property.
  Metric: Primarily subjective, but can be roughly evaluated with things like number of meetings with local government officials and chamber of commerce members, local awards given, positive news stories in the local papers, etc.

- Successfully integrate development activities with cultural resources and environmental policy management requirements in support of the mission of Picatinny.

Metric: Value of resources available for environmental remediation; quantitative measures of environmental impact from tenant activities.

- Provide an opportunity for a reasonable return for the Army and the developer.
  Metric: Total value of leases and developer profit margin.

Potential uses of installation property. Based on an initial, and admittedly cursory, look at the value indicators as applied to Picatinny Arsenal, office-like functions such as back office support for health maintenance organizations, check processing for local banks, and telemarketing operations are all potentially valuable partnering activities. State and local government agencies seeking to expand or relocate their office space may also be potential tenants.

Picatinny's network of transportation resources, proximity to large population centers, and available space make solid and liquid storage attractive. Combining storage space with one or more of the available office buildings also makes extensive shipping/distribution operations for a local retailer or manufacturer an option.

Surprisingly, laboratory and manufacturing facilities appear less valuable because of expected local reductions in these industries. Exceptions here may be R&D partnerships in precision optics, X-ray technology, and paints/coatings. These growing fields appear appropriate to the area and offer potential dual-use technological capabilities for the Army.

Table 2.2 updates Table 2.1, including a column reflecting the applicability of the value indicators to Picatinny.

Choosing Mechanisms to Develop PPPs

Once there is a decision to proceed with a real estate PPP, there is the question of which mechanism, or tool, to use in developing it. The DoD has a number of tools for public-private real estate partnerships; these primarily include leases, facilities-use contracts, and special legislation. Leasing is a general, albeit limited, authority that applies to all Army property. Facility-use contracts are typically used

Table 2.2
Value Indicators for Assessing Potential PPPs: Application to Picatinny

Indicator	Relevant Question(s)	Relevant Variables	Application to Picatinny
Quality of local community	What are the relevant economic and demographic conditions in the community?	Projected economic growth	0.4% local and 11.4% state
		Projected employment growth	27,000 new jobs expected (1998–2006)
		Projected growth industries	Business, health industries
		Projected population growth	8.4% local and 5.4% state
		Projected labor force growth	8.6% local and 7.3% state
		Average education	87% high school; 37% college
		Office/industrial vacancy rates	9.7% office and 6.4% industrial
		Price/square foot of office space	$26/square foot class A space
		Distance to metropolitan area	28 miles west of NYC and 60 miles north of Philadelphia
		Major local university	Drew, Fairleigh/Dickson
		Major local employers	ATT, Lucent, Warner Lambert
		Special development incentives	Designated foreign trade zone; local tax credits
Quality of installation	What are the installation's major land, construction, and infrastructure assets?	Buildings	Over 2.7 million square feet
		Type of service conducted	Science and engineering (S&E)
		Equipment	Lab and engineering
		Available workforce	Over 3,000, mostly for S&E
		Land	Over 6,000 acres
		Utilities	Excellent road, rail, and air transport

Real Estate Public-Private Partnerships 29

Table 2.2—continued

Indicator	Relevant Question(s)	Relevant Variables	Application to Picatinny
Available capacity	How can current installation assets be consolidated, relocated, or restructured to accommodate a PPP?	Buildings	Office, labs, and computing facilities can be consolidated
	What military assets can be used for the PPP?	Dual-use equipment	Office buildings, lab, and warehouse space
Criteria for success	Is the PPP in consonance with the installation's military goals?	Army mission at installation	Home to TACOM-ARDEC; PPP not inconsistent with mission and can assist in meeting noncritical needs
	Are there other benefits to the PPP?		Environmental remediation & revenue
	Does the PPP meet the criteria for success as outlined by the February 1999 GAO report?	Catalyst for change Statutory basis exists Detailed business plan Coherently organized structure Stakeholder support	DoD Directives in place 10 U.S.C. 2667 In progress In progress Local community/private sector involved
Potential uses for installation assets	What types of private enterprise should be encouraged to engage in a PPP?	Qualities of potential businesses should be consistent with installation assets capabilities	Available office space/equipment can fulfill variety of uses; labs useful for private R&D

in specific circumstances and special legislation normally applies to individual programs or installations.

Leases

The leasing authority granted to the Army by 10 U.S.C 2667 allows the Secretary of the Army to lease nonexcess real property to private entities. Traditionally, these leases were used for agricultural and grazing purposes, for antenna sites, and for morale, welfare, and recreation (MWR) facilities. More recently, there have been some efforts to lease facilities on a limited number of Army bases to other types of businesses. To date, however, such efforts have been limited by restrictions associated with the statutory leasing authority (10 U.S.C. 2667) cited above. For example, only 50 percent of rents, when provided in cash, are returned to the leasing Army facility. The remaining 50 percent is used more generally by the Army in support of other facilities. More important, perhaps, Army use of received rent payments was through the standard appropriations process, which effectively diluted any advantage this money would otherwise have given to the Army. Rents, though, could also be provided "in-kind." This means that lessors could provide maintenance, repair, and real property improvement services in lieu of paying cash rent. The limitations on in-kind payments were that they had to be performed on the rented property, and new construction was not permitted. An additional limitation of 10 U.S.C. 2667 was an allowable lease term of only five years, a term length inappropriate for many business leases.

The 2001 defense appropriations bill, H.R. 4205, which was recently enacted, amends 10 U.S.C. 2667 in Section 2812. While hopes that Congress would greatly expand the Army's leasing authority were not met,[15] Congress did somewhat improve the statutory incentives encouraging the Army to lease its underutilized property. Most important, in-kind consideration for leased property can now include new construction. Additionally, in-kind consideration is no longer limited to the leased property, although Congress must be informed if the value of in-kind consideration for a lease exceeds $500,000.

---

[15]For example, the lease period remains limited to five years, and cash receipts must still be deposited in a special account to be spent through the appropriations process.

Facilities-Use Contracting

Facilities-use contracts are an additional tool available to the Army for keeping otherwise underutilized resources productive. Under Federal Acquisition Regulation (FAR) part 45, a private entity contracted to work for the government can, under certain conditions, use government facilities to fulfill contract obligations. The private contractor operating a government facility is also entitled to provide the same facilities, real estate and equipment, to its subcontractors. The Army maintains a level of control over the nature of these subcontracts by requiring its facilities' contractors to obtain Army authorization for each of the subordinate use agreements.

The Army achieves two main benefits from facilities-use contracts: overhead costs are better allocated, thus improving unit price, and maintenance costs are offset or reduced. The private partner may earn income as a result of the work being done at the facility, and may also avoid equipment purchasing costs.

The Anniston Army Depot (ANAD) has successfully employed facilities-use contracts in its partnering activities.[16] Facilities-use contracts are also the contracting mechanism between the Army and the facility contractor in the ARMS program, thus allowing ammunition plant facility contractors to sublet underutilized facilities to other commercial ventures.

Special Legislation

In some cases, the restrictions inherent in 10 U.S.C. 2667 lease legislation have invited direct appeals to Congress. Typically, enthusiasm for PPPs on specified installations is first promoted locally. Once the affected military department, the local political leadership, and the business community come to general agreement about the broad framework of the proposed PPP, specific legislative language is drafted, lobbied for, and introduced into one of the defense funding bills. The advantage to this approach is that it codifies the PPP as a

---

[16]From the Anniston Army Depot Web site: "Facility Use Programs—Agreements wherein public and private entities use ANAD facilities available as underutilized by ANAD operations under the authority of Federal Acquisition Regulation (FSR) Subpart 45 and the Army's Supplement to the FAR." *http://www.anad.army.mil/Partner.htm*.

law and can be as specific as necessary. But special legislation is clearly limited, since it usually applies only to specific installations and programs, and significant political capital must be expended for each PPP advanced in this manner.

Examples of recent special legislation enacted for the benefit of a specific installation are Brooks Air Force Base in Texas and the Ford Island complex at Pearl Harbor, Hawaii. The legislation in these cases authorized the development of installation property according to a plan introduced as part of the legislation. Each of these programs was promoted locally and through the affected congressional delegations.

The tools available to the Army for the implementation of real property PPPs are limited. Despite this, there are examples demonstrating that these tools, used diligently and imaginatively, offer opportunities to create more PPPs with Army real property.

PPPs IN CONCLUSION

Absent a very significant Base Realignment and Closure (BRAC) round, the U.S. Army is likely to maintain real property that is excess to its current needs and its current maintenance abilities. Since such a BRAC is highly unlikely, it is important that the Army find ways to manage and maintain the property that it will continue to hold. PPPs are one means of putting the Army's property to good use and in a manner that provides the additional resources necessary for the effective stewardship of the vast holdings entrusted to the Army. Though there are a number of significant issues associated with PPPs on military property, we believe that all of them are solvable. More important, making the establishment of PPPs Army policy, rather than the exception, should greatly expand both the quality and quantity of PPP proposals.

Chapter Three
VENTURE CAPITAL APPROACHES

"C.I.A. Tries Foray Into Capitalism:
Sets Up Venture Capital Fund Concern to Back High-Tech Projects"

—John Markoff
The New York Times
September 29, 1999

CONTEXT

Over the past decade, the amount of resources the Army devotes to research and technology development has stagnated. For example, in terms of total research, development, test, and engineering (RDT&E) (budget categories 6.1–6.7), the Army's funding has declined from about $6.8 billion in 1993 to about $5.2 billion in 2001; in addition, funding for total S&T (6.1–6.3) has dropped from about $2 billion in 1993 to about $1.3 billion in 2001.[1]

Despite this stagnation in research and technology development, the Army's performance expectations for new and future equipment, and hence their technical content, continue to grow. Specifically, S&T capability is postulated as a central driver in the Army's planned transformation. In particular, designing the core of the Army's transformation Army—the Future Combat Systems (FCS)—will require implementing significant S&T advances.

---

[1] Budget categories are defined as follows: 6.1, Basic Research; 6.2, Applied Research; 6.3, Advance Technology Development; 6.4, Demonstration and Validation; 6.5, Engineering and Manufacturing Development; 6.6, RDT&E Management; 6.7, Operational System Development.

Given this asymmetry between declining resources and increasing needs, developers of Army materiel are forced to look elsewhere for their technology. "Elsewhere" in this case is the commercial technology sector, upon which the Army greatly depends. Unlike the Army (or for that matter DoD), the commercial sector, spurred by competitive forces and a population that eagerly accepts new technologies, has seen its R&D spending quadruple in three decades and continues to grow at more than 4.5 percent per year.

Contrary to popular opinion, commercial R&D spending is not done solely at the product development stage. Private companies have outspent the federal government in applied research for a number of years now and are spending a large and growing percentage of the country's basic research dollars.[2] What this means for the Army is that a growing portion of the technical innovation occurring in the country is happening in the commercial sector, thus making Army access to that sector more important than ever.[3]

In fact, going back to the FCS example mentioned earlier, many, if not most, of the required technologies are of commercial interest and are being actively pursued in the commercial technology sector. This is the finding of both a RAND study and other Army studies.[4]

---

[2] The Industrial Research Institute has concluded that $10.9 billion of industry's $166 billion in R&D spending in 1999 was devoted to basic research—the "blue-sky" pursuit of products and services that might lie a decade in the future (Larson, 1999). Also see National Science Foundation (1999) for detailed statistics.

[3] One quantitative measure of innovation is patenting activity. Data available through the U.S. Patent and Trademark Office closely mirror the trends seen in resource allocation for R&D. The number of commercial patent applications has doubled in the last decade (U.S. Patent and Trademark Office, *TAF Special Report*, 1999), while the number of Army patent applications has stagnated (U.S. Patent and Trademark Office, *TAF Profile Report*, 1999).

[4] In the summer of 1998, the Army conducted a seminar war game at the Army War College. The purpose of the seminar was to map the future Army's technology requirements. One of the seminar's products was a listing of technologies that the Army considered necessary to "enable" the future doctrine and tactics then being experimented with (Lavine, 1999). We estimate that fully 75 percent of the listed technologies have significant commercial application and content. Likewise, the Army Science Board noted that many of the FCS technologies mature in time to meet the FCS fielding schedule because of the ability to leverage the commercial sector (Burger, 2000).

The fact that so many FCS technologies will be developed first in the commercial sector can be quite advantageous for the Army. But like most DoD organizations, the Army has difficulty gaining and maintaining access to the advanced technology being developed in the commercial sector. For a variety of legal and cultural reasons, the Army seems to prefer to rely on its traditional suppliers, and many of the companies working on advanced technology seem to avoid contracting with the DoD in general.

In this chapter we examine another option for the Army to pursue in gaining and maintaining access to the advanced technology being developed in the commercial sector: establishing a venture capital fund, beginning with a discussion of why the Army has difficulties accessing the commercial technology sector and what it has done to address these difficulties, then turning to why a venture capital fund makes sense, and concluding with an implementation strategy for such a fund.

WHY DOES THE ARMY HAVE DIFFICULTY ACCESSING THE COMMERCIAL TECHNOLOGY SECTOR?

There are a number of reasons the commercial technology sector has proved a difficult target for the Army. Some of the blame can be assigned to the consolidation of the defense industrial base that has occurred over the last decade. Many companies that had traditionally done business with the government, but also had significant commercial business, spun off their military businesses. These businesses were either acquired by the remaining military prime contractors or, in some cases, became independent companies that primarily focused on the military market.[5] This consolidation tended to increase the isolated nature of military R&D. However, defense consolidation may be more symptomatic than causal.

In terms of the causal reasons, there are real and perceived obstacles that prevent more collaboration between the DoD and commercial

---

[5]General Electric, FMC, IBM, Dupont, and Honeywell are prominent examples of companies that have divested all, or most, of their defense businesses.

companies that perform research and product development.[6] The most prominent barriers to greater collaboration are (1) intellectual property concerns, which combines with the fact that most companies do research for their own purposes, not as a service for hire; and (2) excessively bureaucratic requirements and the related distrust of government involvement and oversight in company affairs. When commercially oriented companies weigh these burdens against the relatively small size of the Army market, other limitations on profits, and the perceived fickleness of the government as a customer, the benefits of collaboration generally fail to overcome them.[7]

As for intellectual property rights, innovation is almost by definition an intellectual endeavor. Making a profit on an innovative idea requires monopolizing it to a degree and getting it to market before or more successfully than competitors do. A company that relies on its intellectual property to gain a competitive advantage will thus be very reluctant to cede that property. There is concern among potential industry partners that the government just does not understand or sympathize with this (Chen, 1999). This means that contracting for R&D with most commercial firms will be difficult, since traditional government contracting precludes flexibility in the area of intellectual property.

The stigma of government bureaucratic "red tape" is also a problem. Because the government procurement system is so different, government contracting requirements often force companies to create positions and business units solely dedicated to dealing with it. Setting up separate business units or divesting defense business is also sometimes seen as a necessary shield against excessive government oversight of the entire company. The thought of granting government auditors/inspectors access to company records is a strong deterrent to working with the government, yet it is difficult for the

---

[6]Stan Z. Soloway, Deputy Under Secretary of Defense (Acquisition Reform), noted in a recent *Wall Street Journal* article that three-quarters of the top information-technology companies will just not do research for the DoD (Chen, 1999).

[7]During a conference entitled "Perspectives on Other Transactions," there was much discussion about why nontraditional military suppliers (NTMSs) continued to be reluctant to do business with the government, even when OT authority was available. The listed reasons were identified as the most important culprits (Held, 1999).

government to accept normal business relationships in the matter of oversight.[8]

The obstacles noted above are most relevant to the traditional tools used by the military to contract for research: grants and FAR-type contracts. Both of these are saddled with substantial regulations that limit their flexibility in many areas important to commercial businesses, such as accounting procedures, intellectual property rights, and oversight. Previous RAND research verified that these limitations have turned away many nontraditional military suppliers (NTMSs) (Horn et al., 1997). Additionally, since the DoD provides research grants only to educational and other nonprofit organizations, the utility of grants in accessing the commercial base is circumscribed.

WHAT HAS DoD DONE TO ADDRESS THESE DIFFICULTIES?

These difficulties are not unique to the Army—they are DoD-wide concerns. In this section, we discuss three of the solutions DoD has pursued to overcome these difficulties and illustrate the relative ineffectiveness of each.

New Tools Designed to Access the Commercial Technology Base

Recognizing the limitations inherent in its traditional contracting tools, DoD has gained a number of new contracting tools designed specifically to access the commercial technology R&D base. Unfortunately, their success in attracting NTMSs has been limited. The most important of the new tools is the Other Transaction (OT), codified in 10 U.S.C 2371. The statutory language authorizing this contracting method provides a great deal of flexibility, because it defines OTs in the negative by stating that they are not contracts, grants, or

---

[8]For example, during a conference, one government auditor expressed an opinion that no company should fear opening its books to government auditors, provided they were doing everything correctly and legally. Another related a story about one company that had an OT agreement with the government. The agreement specifically limited the government's oversight authority. The auditor expressed some surprise that when he showed up at the company's door and asked to examine the company's records, he was rebuffed (Held, 1999).

cooperative agreements (CAs). The practical result of this negative definition is that the regulations governing the traditional contracting tools do not apply to OTs. Intellectual property rights, government oversight, cost-sharing, and business arrangements are all negotiable. In fact, by a plain language reading, it would seem that OT legislation allows any kind of agreement to conduct research between the government and a contractor, provided the agreement is in the government's interest. Thus, it would seem that the government should be routinely able to establish "business-like" arrangements with commercial businesses for research collaboration through the use of OTs.

Unfortunately, the Army's record of using OTs is not impressive. Since getting the authority to use them, the Army has used OTs on less than one-quarter of 1 percent of its research contract actions.[9] The DoD Inspector General has also noted that most of the OTs signed so far have been with traditional military suppliers, despite the fact that OTs were meant to facilitate agreements with NTMSs (Office of the Inspector General, 1999).[10] It is unclear why the Army has been so reluctant to use the OT authority. Likely reasons include the Army procurement culture, lack of training, and a general difficulty on the part of the Army in finding and marketing to NTMSs who are willing to collaborate on research projects.

Like the OT, the CA was established, in part, to make R&D contracting more flexible and more attractive to the commercial sector. CAs have more restrictions than OTs; not surprisingly, they have not been

---

[9]The Army signed 48 OTs for research and prototypes from 1996 through 1998 (Office of the Inspector General, 1999). The Defense Contract Action Data files indicate a total of over 21,000 Army contract actions for R&D during the same period. The DoD Inspector General's report also provides a breakdown indicating that 13 of the OTs were for prototypes, while 35 were for research. Tellingly, 10 of the 13 prototype OTs are part of DoD's Commercial Operations Support Savings Initiative (COSSI) program, which *mandates* the use of OTs (U.S. Department of Defense, *Report on Other Transaction Awards for Prototype Projects*, 1999).

[10]For example, an examination of the 24 OTs for research that the Army signed in 1998 indicates that they involved 30 non-Army participants. The companies' Internet descriptions of themselves revealed that only 17 percent of participants were commercial companies that were NTMSs. These companies accounted for less than 10 percent of the Army OT for research contract dollars awarded.

particularly successful in attracting commercial-sector companies into collaborative or contractual R&D relationships with the Army.[11]

The Small Business Innovative Research (SBIR) and the Small Business Technology Transfer (STTR) programs are relatively successful in terms of funding a large number of small businesses interested in conducting research for the Army.[12] However, like most other government contracting programs, the SBIR and STTR programs tend to attract companies that are comfortable doing business with the government and that have the government as their most significant customer.[13] While these programs do contract with a large number of start-ups and other small businesses that do not have experience with the government, limitations in the structure of the SBIR/STTR program curb its utility vis-à-vis these companies.[14] As a result, commercial success has not been a hallmark of the technologies flowing out of the DoD's SBIR program.[15]

The "Fast Track" SBIR program offers additional incentives to small businesses that find additional funding sources. The companies

---

[11] For example, the Army Research Laboratory's heralded Federated Laboratory Concept used the CA as a funding mechanism to establish three consortia for different research areas. Traditional military contractors head all three consortia (Brown, 1998).

[12] The SBIR and STTR programs funded about 300 programs and had a budget of $110 million in FY99 (Army Research Office, Business Opportunities Web site, *http://www.aro.army.mil*).

[13] A random sampling of the Army's FY99 and FY00 Phase 1 awardees indicates that 50–75 percent of the awardees are government contractors and at least 60 percent have received multiple awards. Likewise, the GAO found that two-thirds of the companies receiving awards had received earlier awards and that just 25 companies had received 11 percent of the total contract dollars from 1983 to 1987. During this time 45,000 SBIR contracts were awarded (U.S. GAO, 1999a).

[14] By policy, the DoD SBIR program uses traditional government contracting as the agreement form. Though some of the contracting rules are relaxed for small businesses, many of the factors that make government contracting unappealing to commercial businesses remain. There are also significant funding limitations. (The maximum funding for SBIR is $850,000, though most programs receive less. The Fast Track program may provide some additional funding.)

[15] A 1996 survey found that 75 percent of the commercial sales related to technologies developed under DoD's SBIR program were developed in just 4 percent of the projects. Commercialization is becoming a more important evaluation criterion, but it is too early to tell whether the added emphasis will improve the commercialization of SBIR technologies (U.S. GAO, 1999a).

participating in the Fast Track program appear to be having greater commercialization success,[16] which suggests potential opportunities to improve the SBIR program with the venture capital approach discussed in the next section.

Funding Specifically Aimed at Commercial Technology

Recognizing the potential of commercial technologies, the DoD has established a number of programs whose funding is specifically targeted toward increasing the commercial technical content of DoD research and equipment. Success in attracting commercially oriented firms is mixed. The Commercial Operations Support Savings Initiative (COSSI) is a funded program whose goal is to reduce operations and support costs by putting commercial technology into fielded systems. A briefing describing the Army's program provided four program examples. In three of the four programs, the contractor was a traditional military supplier (Rohde, n.d.). The Dual Use Science and Technology (DUST) program funds R&D programs that have both military and commercial application. In 1999 the Army awarded 27 DUST contracts.[17] Nineteen of the contracts went to traditional military suppliers, six went to NTMSs (although a traditional military supplier subsequently acquired one of them), and two went to educational institutions (Dual Use Science and Technology Program, 1999). The mix changed in FY00, with 7 out of 11 contracts going to NTMSs, indicating improvement, but funding remained limited despite previously anticipated increases. Importantly, both the COSSI and DUST programs are being terminated at the end of the current year.

HOW DOES A VENTURE CAPITAL FUND WORK?

There is certainly no lack of recognition among the Army leadership of the necessity of working with the commercial technology sector. As noted above, though, developing the collaborative ties between

---

[16]Discussions with Robert S. Rohde, Deputy Director for Laboratory Management, Office of the Assistant Secretary of the Army (Acquisition, Logistics, and Technology), October through December 2000.

[17]DUST and COSSI contracts are required to be either CAs or OTs.

the Army's R&D community and commercial technology developers remains difficult. The old contracting tools are inappropriate, and Army contracting officers and the government oversight community, probably through a lack of training, resources, and authority to use it, appear reluctant to use the OT. As a result, Army penetration of the commercial technology sector proceeds at an indifferent pace. We suggest that one solution is to sidestep the barriers by establishing a venture capital fund for technology development. In this section, we describe how such a venture capital fund would work, while the next section describes the benefits such a fund would have for the Army.

Venture capital describes a broad category of investment that has two defining characteristics. First, venture capital investments are made in businesses that have a high risk of failure but also potentially high returns. Second, venture capital investments are accompanied by a fairly high degree of investor involvement in the investee firms (Gladstone, 1988, p. 3). Although venture capital is normally associated with equity capital in which funding is provided in exchange for an equity stake in the business, not all venture capital is equity capital. Other investment mechanisms, such as royalties on future profits or high-risk, high-interest loans, also fall into the category of venture capital.

Venture capital funds are organized in a number of ways: as limited and general partnerships; as public, private, and limited liability corporations; and as subsidiaries to larger companies. However, the traditional organizational structure is the limited partnership. This structure is organized around a general partner and one or more limited partners. The general partner runs the fund and is fiducially responsible to the limited partners. The general partner may invest some of his own capital into the fund but is generally compensated with a management fee and a percentage of the fund's total return. The limited partners are passive investors with limited liability who place their investments in the hands of the general partner with the expectation that they will earn a good return. Importantly, they avoid active involvement in fund management.

Once funding has been raised, the management of a venture capital fund goes about the process of evaluating investment opportunities and selecting companies for funding. A primary function of the venture capitalist is to gather information about potential markets, technical feasibility, competition, and other facts that will impact the probability that a new business will succeed. This knowledge comes from a number of sources, including past experience, contacts in the market segment, other venture capitalists, trade journals, and the business plans submitted by entrepreneurs looking for funding. Using a combination of experience, analysis, advice, and intuition the venture capitalist decides which ventures to fund and the extent to which they will be funded.[18] As mentioned above, venture capital involves various funding mechanisms, though equity financing is the most common.

In general, venture capitalists fund relatively new and rapidly growing companies. There are a couple of reasons for this. First, newer companies tend to be more efficient in that they have much less overhead and a core staff more directly affected by the success of the company. Second, and perhaps more important, new and growing companies have a greater potential for the high rates of return that venture capitalists require.

Venture capitalists prefer to fund innovative services and products because they offer significant returns through developed markets in a relatively short time frame.[19] Innovative products and services are developed from relatively mature but unexploited ideas and technologies. In other words, venture capitalists do not normally invest in basic research, although they often take the fruits of basic research and turn them into something useful and marketable. They also look for new uses or combination of existing technologies. For example, e-commerce combined Internet technology with electronic fund transfer and modern inventory and distribution practices to create a new industry.

Once funding is extended, the venture capitalist usually takes an active role in building the investee company. It is this role, along

---

[18]Traditionally, venture capitalists relied on requests for funding from entrepreneurs to identify potential investment opportunities. That may be changing now. Gilman Louie, the CEO of In-Q-Tel, told us in an interview that more venture capitalists are creating investment opportunities themselves by identifying potential market niches and creating companies from scratch to fill those niches. This model may be more appropriate for an Army venture capital fund.

[19]Gladstone also discusses evolutionary and substitute products and services, but we omit these, since they are not germane to the discussion. (Gladstone, 1988, p. 3).

with the risky nature of the investment, that defines venture capital. Typically, the venture capitalist will provide a number of services. These include continued refinement of the business plan, help with putting together a management team, valuable business contacts, assistance in securing additional funding, management assistance, and help in marketing (Sargari, 1992, p. 7-8).

WHAT ARE THE BENEFITS TO THE ARMY OF ESTABLISHING A VENTURE CAPITAL FUND?

As mentioned above, we recommend that the Army bypass the obstacles it faces when accessing the commercial technology sector by setting up a venture capital fund like the model just described that invests in companies and technologies that are of interest to the Army and that have potential for significant commercial market penetration in the longer term. In this section, we examine some of the benefits that the Army would gain by establishing a venture capital fund.

Can Exploit Innovation

Though relatively young in its current forms, venture capital has been extremely successful in developing and exploiting innovation. Many of the most inventive companies in the world, including Intel, DEC, Apple, Microsoft, Sun Microsystems, FedEx, Genentech, and Netscape, used venture capital as a key resource and are examples of its success. Empirical evidence also supports the claim that venture capital spurs innovation. Although difficult to measure directly, several studies have noted the positive correlation between the use of venture capital as a funding tool and indirect measures of innovation. One examination found that venture-backed companies spend 45 percent of their equity on R&D in their first five years (PriceWaterhouseCoopers, 1998).[20] Similarly, R&D expenditures in Europe represented, on average, 8.6 percent of total sales for venture capital–backed companies compared to 1.3 percent for the "blue chip" European companies (European Private Equity and Venture

---

[20]By way of comparison, manufacturing industries in the United States spend just over 3 percent of sales on R&D (Wolfe, 1999).

Capital Association, 1999). Using patenting of inventions as an indicator of innovation, a significant study suggested that a dollar of venture capital was 5–14 times more effective than a corporate R&D dollar in terms of innovation (Kortum and Lerner, 1998).

The reasons for venture capital's success are its inherent incentives and an organizational structure that facilitates the development of innovative ideas. Young, small, and growth-oriented companies typify the investee. Their potential products or services are new and intended to develop new markets or redefine older ones. The company founders are risk takers motivated by their vision. Investors are experienced businessmen and women who are also risk takers but who expect to be well rewarded for taking those risks. They are adept at managing young companies and commit, in addition to funding, significant intellectual capital, business experience, and time to the companies they back to maximize the opportunities for success.

Can Be Used by Public and Large Private Organizations for Technology, Investment, and Nonfinancial Reasons

Many large corporations, even those with substantial internal R&D capabilities, recognize how well venture capital exploits innovation and now use it to develop technologies for their businesses. In one example, Xerox Corp. put together a successful venture capital fund to turn Xerox-developed technologies, otherwise dormant, into marketable products (Gompers and Lerner, 1998). In addition, Microsoft has earned the reputation as an acquirer of new, venture-backed start-ups that can contribute to its key technologies, and Lucent Technologies has a $100 million venture capital fund that it uses to invest in new technologies, despite its in-house staff of 30,000 scientists (Taptich, 1998). These examples are important when considering whether a venture capital model will work for the Army. They suggest that large organizations, even those with organic R&D capabilities, have found venture capital to be an efficient use of limited R&D resources.

Beyond these private-sector examples, use of venture capital is also successfully spreading to the public sector. A number of state governments have set up successful venture capital funds for a variety of reasons, such as job growth, expansion of light industry, and the development of companies that correct perceived problems (e.g., environmental).[21] Financial return on investment (ROI) is typically a secondary motive for state funds. At the federal level, the Department of Energy's Argonne National Laboratory, which is run by the University of Chicago, has had a special relationship through the ARCH Development Corporation with the ARCH Venture Fund. This relationship helps commercialize the discoveries made at Argonne's laboratory facility, thus providing the public greater access to technologies funded with Department of Energy money (ARCH Development Corporation and the ARCH Venture Partner's Web sites, 1999).

Perhaps the example closest to that envisioned for the Army is the CIA's In-Q-Tel enterprise.[22] The CIA recently established In-Q-Tel to solve some of its most difficult information technology problems, and venture capital is one of the tools In-Q-Tel uses to address the CIA's technology needs. In-Q-Tel has been in existence for only about a year and a half, not long enough to determine its ultimate financial success, but it appears to have made a very promising start in terms of technology development. As of this writing, In-Q-Tel already has, or will have very shortly, product solutions to some of the requirements presented by its sponsor, the CIA.

Can Better Access Commercial Technology

A venture capital organization funded and chartered by the Army but run outside the government by a venture capital professional could circumvent many of the obstacles that prevent greater collaboration

---

[21]The Massachusetts Technology Development Corporation (MTDC) is a good example. It was established by state law in 1980. In its 20 years of existence the MTDC has invested nearly $40 million and is now self-sustaining, relying on returns from earlier investments to fund new investments. The MTDC Web site lists four primary objectives for the fund: (1) to help create primary employment in Massachusetts; (2) to attract and leverage private investment in Massachusetts companies; (3) to foster the application of technological innovations where Massachusetts companies are, or can be, market leaders; and (4) to nurture entrepreneurship among Massachusetts citizens, planting the seeds for long-term economic development in the Commonwealth.

[22]The authors have interviewed both CIA and In-Q-Tel staff on several occasions. Many of the concepts presented in this section are borrowed from the CIA's establishment and development of In-Q-Tel.

between the Army and the commercial technology sector. In this scheme, the Army venture capitalist would act as a middleman who understands the needs of the business and technology communities and who can shape agreements that meet those needs and that also solve Army technology problems. Since the venture capital organization would be outside the Army, it should be better able to gain the trust of commercial clients and also act more quickly and flexibly than could the Army's current contracting organizations.

Can Leverage Non-Army Resources

Another important reason for the Army to develop a venture capital fund is the way in which it can be used to leverage non-Army resources. Today, most Army research is conducted exclusively with Army resources. While some of the newer contracting tools allow cost sharing on research projects, there are practical and legal limitations to the amount of cost sharing available. In contrast, venture capitalists and the entrepreneurs they support freely seek funding from any number of sources. Assuming the Army's fund invests in technologies that also have considerable commercial potential, significant outside co-investment is quite possible and likely. The advantages are obvious. Leveraging allows the Army to stretch its own R&D resources so it can accelerate the development of key technologies while continuing to invest in a diverse range of new ideas.

Can Provide a Return on Investment

Commercial venture capital's reason for being is to earn a ROI. As mentioned earlier, venture capitalists expect large returns in compensation for the risks they place on their investments. Identifying an average return across the venture capital industry has proved difficult, and estimates vary considerably. Despite this, the success of the venture capital industry is clearly implied by the associated exponential growth in investment, as shown in Figure 3.1 (Gompers and Lerner, 1999, and VentureOne Corporation Web site, various years).

Most of the technologies appropriate for investment by an Army venture capital fund will be those that have a near-term Army requirement but a longer-term commercial potential. By using a venture

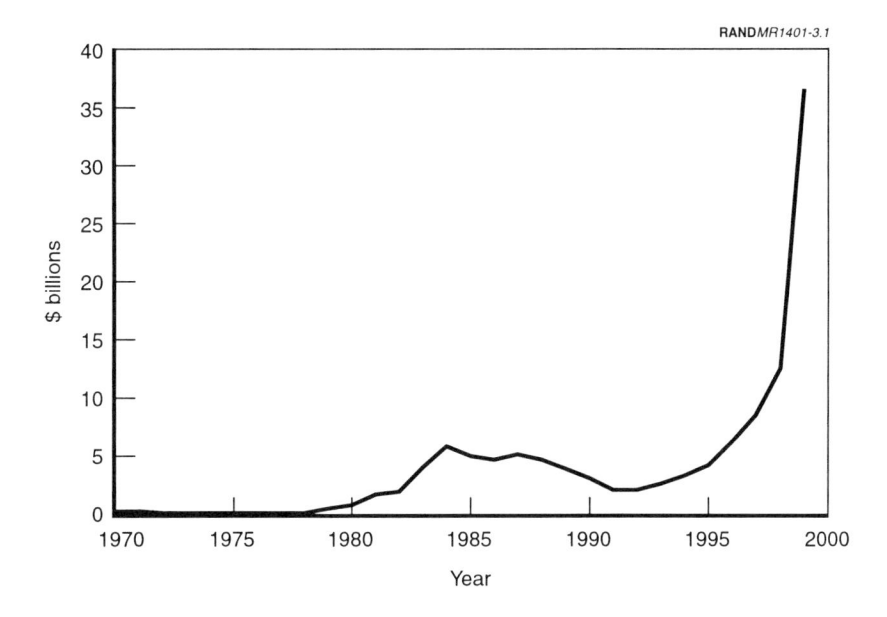

Figure 3.1—Venture Capital Funds Raised

capital model to make the initial investments in new technologies, the Army will be able earn a ROI as the commercial market for these technologies grows. This return can then be used to strengthen Army R&D further through reinvestment by the Army's venture capitalist.

Can Give Rise to Entire Industries

There is a general rule of thumb that radically new technologies are usually developed, marketed, and matured by new companies. With some exceptions, making bold technological and product line shifts is difficult for established companies, which usually prefer to evolve along the established lines that have been successful for them in the past. In its short history, venture capital has thus become the source of start-up money for many emerging industries. In the military, many of the transforming technologies also spawned new industries. Repeating rifles, radio, aircraft, and (today) the integrated circuit come readily to mind. Although these examples eventually grew very large commercial markets, the military was generally the first customer and, thus, was largely able to guide the development of the industries and technologies involved. With most R&D occurring in the commercial sector today and with the change in markets, many of tomorrow's transforming technologies (e.g., biotechnology and networking) are being developed with little input from the military. By creating its own venture capital fund, the Army can regain some of its access and influence in emerging industries.

POSSIBLE IMPLEMENTATION STRATEGY FOR AN ARMY VENTURE CAPITAL FUND

In this section, we discuss how the Army might implement a venture capital fund, starting with how such a fund might be established, moving on to how it might be run and discussing what technologies are appropriate for such a fund, and concluding with how such a fund would be integrated with other Army technology programs.

Establishing an Army Venture Capital Fund

For this report, we name the Army venture capital fund the Army Innovation Investment Corporation (AIIC). The AIIC could be formed under two alternative scenarios. Under the first, the Army would form an agreement with an existing organization, such as an existing venture capital fund or a federally funded research and development center. This agreement would be a research agreement with a set of Army problems to be solved, rather than an agreement to establish a venture capital fund. However, by using the existing, and very flexible, authority of 10 U.S.C. 2371, the OT, the agreement could be arranged such that a venture capital fund would be, at the least, one of the tools used to solve the Army's problems. The agreement could also lay out various other details, such as how returns on investments are handled, how assets are distributed in the case of agreement termination, and how intellectual property rights are handled. At some point, after the initial agreement, we envision that the AIIC would be divested from the parent organization and could be run independently along the lines we recommend below. The advantage to this approach is that the Army has the authority to do it now, provided that the funding is made available. The disadvantage is that the OT has never been used in this manner. Given the

Army's hesitancy to use the OT at all, it is probable that without very high-level insistence, the Army's procurement establishment will be extremely reluctant to endorse this kind of arrangement and use of the OT.

Under the second scenario, specific authorizing language and the funds to set up the corporation would be included in the DoD's annual authorization and appropriations process. The advantage to this approach is that the establishment of AIIC would carry the unambiguous legal authority conferred on it by Congress and codified in public law. There are several potential disadvantages. The first one is the possibility that the Army could lose control of the process. What we mean is that once Congress begins the process of writing the laws allowing the formation of the company, it could do so in a way that is not aligned with Army concepts. In such a case, Congress, not the Army or the founders of AIIC, would control the formation of the company. Another disadvantage of this scenario is the time required and political capital needed to advance the concept through the appropriate congressional committees. The CIA was able to persuade Congress to fund In-Q-Tel very quickly (a matter of months), but In-Q-Tel was conceived of at the highest levels of the CIA, so the required high-level support and lobbying was ensured. The Army's problem is more complicated because committed, high-level support for the idea must first be developed within the Army. Further complicating the Army's task is that unlike the CIA, the Army has another layer of authority, the DoD, between it and Congress.

Perhaps the best way to establish AIIC would combine the two methods. Under a combined process, the Army would begin in a small way through the OT method described above. For a relatively modest amount of money, perhaps less than $10 million, the Army would partner through an OT agreement with an established organization to begin work on a limited set of problems. (The type of problem appropriate for an Army venture capital fund will be further described below.) The Army partner would organize and staff itself, if not already set up as such, to use venture capital as a tool for solving the problems in the partnership agreement. With an agreement in place and a small number of projects under way, the Army could then look for congressional endorsement and additional funding through the authorization and appropriations process. By putting its venture together in this manner, the Army will retain significantly more control over the shape of AIIC, the structure of the Army's relationship with it, and the timing of its establishment. Also, more time will be available to build congressional support, and the lessons learned during the establishment of AIIC can be incorporated into any statutory language that might emerge.

One Possible Model for an Army Venture Capital Fund

The AIIC will, necessarily, be run as a nonprofit corporation. This means that any income generated through its investments will be reinvested to solve more Army problems or will be returned to the U.S. Treasury. Maintaining a nonprofit status eases tax issues that may surface but, more importantly, eliminates the appearance of impropriety that could arise in a government-sponsored, for-profit organization.

We recommend that the AIIC be managed by a board of directors made up of private citizens selected for staggered two- to three-year terms. Since the Army is the ultimate "stockholder" of AIIC, selection of board members should be at least partially controlled by the Secretary of the Army or his staff. For instance, selection of candidate board members could be done by a committee made up of board members not due for replacement or retention and senior Army staff at the Assistant Secretary level. Candidate board members would then be confirmed by the Secretary of the Army.

Having the right mix of the right people will be the most important factor in forming and subsequently maintaining AIIC. The staff must contain a mix of personnel with business, technology, and government experience.[23] This eclectic group must then be integrated by a strong leader who not only understands the complexity of technologically oriented business deals but who can also navigate the political and bureaucratic terrain inherent in an organization of this type. We recommend that the board of directors select the chief executive officer (CEO), who becomes an automatic board member. The

---

[23]In the case of In-Q-Tel, this requirement has been satisfied by mixing very bright young people, recently graduated with advanced degrees in business and science/engineering, with personnel experienced in venture capital, business, and government.

board, including the new CEO, then select other key members of the AIIC's management staff, such as the chief financial officer and chief technology officer. This management team, in turn, fills out the lower tiers of the organization.

Getting the correct personnel for AIIC also means establishing adequate incentive packages. AIIC's nonprofit status, the need to avoid conflicts of interest, and the potential for improper appearances imply that employees will probably not be able to accept stock options in their investee companies. Instead, the incentive package will need to be cash based and should carefully balance the requirement to solve Army problems with the potentially conflicting need for AIIC financial success. As an example, employee compensation could include a base salary and a bonus with a potential maximum that is some multiple of the base salary. That multiple would then be based on a combination of portfolio financial performance and the degree to which the AIIC investments solve Army problems.

An Army Advisory Committee composed of personnel from the Army Materiel Command (AMC), the Office of the Assistant Secretary of the Army for Acquisition, Logistics, and Technology (ASA(ALT)), and the Training and Doctrine Command (TRADOC) could form the interface between the Army and the AIIC. The committee would be responsible for communicating the Army's operational requirements and technical needs to the AIIC, which, however, would make all business decisions about investments.

The Advisory Committee would also be responsible for the transfer of technical information from the AIIC back to the Army. Technology transfer is likely to be a more involved process than just a sharing of information through the Advisory Committee. We expect that one of the major responsibilities of AIIC management will be to promote, within the Army, technologies and products being developed by AIIC investments. Successful promotion is critical for two reasons. Primarily, it ensures that Army investments are indeed solving Army problems. Secondarily, we expect that in many cases, the Army will be the first major market for many of the products, so successful promotion can help ensure the commercial and, hence, the ultimate financial success of AIIC investments.

Appropriate Technologies for an Army Venture Capital Fund

Clearly, not every military technology program would be appropriately funded through venture capital. The technology investments suitable for Army venture capital funding would have three prime characteristics. First, the technology must have clear military and commercial applicability. Second, the Army must be in a "power user" position.[24] A power user is one who has a requirement for a new product or technology ahead of other potential users. Because of this position, the power user is normally willing to invest earlier and with a little more risk. Finally, the technology must be "mature enough" to develop into a product or proprietary technology in the limited time and with the limited dollars that venture capital investing implies.

Selecting the correct technology areas for investment will be one of the first and most critical responsibilities of a venture capital team funded by the Army. An example of the kind of technologies we envision as appropriate for Army venture capital are those required for the 21st Century Land Warrior program. For example, lightweight batteries with very long life are required as a power source. Technology that addresses this problem clearly has significant commercial application, while the Army's ambitious requirements place it in a power user position relative to the technology. Whether there is a commercial base of research and development to support Army venture capital investment into one or more technical solutions for Land Warrior would still need to be addressed by the Army's VC.[25]

Integration with Other Army Technology Programs

Finally we note that an AIIC should not be developed in isolation, particularly in light of an Army R&D budget of about $5 billion a year, of which about a quarter is spent on S&T. An Army venture capital fund would need to be fully integrated into this already large effort.

---

[24]"Power user" is a term coined by Gilman Louie of In-Q-Tel.

[25]The Army has in fact borrowed heavily from the commercial sector to address the issues raised by the GAO. Specifically, the weight and electromagnetic interference issues have used commercial technologies that alleviate many of the GAO's concerns (Cox, 2000).

As mentioned already, finding "sponsors" and users for the venture-backed technologies will be critical. Additionally, we see other ways that venture capital could be integrated into existing programs to make them better. For example, the Army SBIR program spends over $100 million per year in what is essentially "seed" money. In other words, the SBIR program funds hundreds of technologies each year that are usually too immature for venture capitalists. While commercialization is a priority of the SBIR program, we believe the program's structure does not support commercialization particularly well. An integrated venture capital approach could help solve this problem by providing the funding and support needed beyond that provided by the SBIR program. Likewise, the SBIR program could be a source of technologies for the venture capital fund, particularly if some of the SBIR awards are given with the AIIC's problem set in mind. Other Army programs could similarly benefit from an Army venture capital if well integrated into a broad S&T program.

Chapter Four

SPINNING OFF ARMY ACTIVITIES INTO FEDERAL GOVERNMENT CORPORATIONS

> Experience indicates that the corporation form of government is peculiarly adapted to the administration of governmental programs which are predominately of a commercial nature, are at least potentially self-sustaining, and involve a large number of business-type transactions with the public.
>
> —Harry S. Truman
> Budget Message to Congress, 1948

CONTEXT

Mergers, acquisitions, subsidiaries, and asset sale are standard tools for the implementation of corporate strategy. In the decade of the 1980s when the corporate strategy of General Motors (GM) involved pulling critical business functions as close as possible under the corporate umbrella, Electronic Data Systems (EDS) was acquired because it was providing all the information, computer, and data services for GM. Moreover, GM was the biggest single customer of EDS. More recently, in 1996, under a new corporate strategy involving a focus on core businesses, EDS was divested in an asset sale with GM retaining some of the stock ownership through a GM pension plan. During each of these eras the GM leadership felt the key to competitive advantage was aligned in directions indicated by the acquisitions and divestments of the time.

The United States has a similar instrument for the implementation of national policy: the Federal Government Corporation (FGC) (U.S.

GAO, 1995; Froomkin, 1995). The first Government Corporation predates the Constitution. The Continental Congress decided in 1781 that a bank owned by our country rather than the only other choice at the time, the Bank of Britain, would better handle the finances of the new nation. The Bank of North America was chartered as a result (Lewis, 1882). The first Federal Government Corporation was the Bank of the United States, chartered in 1791. During the 20th century the FGC has become a common instrument of national policy. Since World War II the Congress has created about one FGC per year, resulting in about 60 in existence today.

A good example of the formation of an FGC is in the area of uranium enrichment. When nuclear testing ceased, the Strategic Arms Limitation Talks (START) were continuing, and a Comprehensive Test Ban Treaty was anticipated, the strategic position that the U.S. government played as a customer in the market for enriched uranium changed. Hynes, Kirby, and Sloan (2000) have traced the various developments leading up to the final decision to spin off the uranium enrichment activity. After several years of consideration, Congress passed the Energy Policy Act of 1992. As part of this act, the United States Enrichment Corporation (USEC) was formed as a wholly owned government corporation. Eventually, the USEC, operating as a government corporation, proved to be successful, and in 1998 it was privatized and its stock began to be traded on the New York Stock Exchange. The initial public offering agreement contained provisions that allowed the government to benefit from any windfalls in profits from the new private corporation.

During the first half of the 20th century, FGCs were a common instrument of national military strategy to capture the manufacturing efficiencies of the U.S. economy for both the execution of and preparations for the two world wars (Lilienthal and Marquis, 1941). In the current era the FGC can be an instrument of national military strategy. In an era of decreasing federal budgets, increased constraints on personnel, and growing emphasis on achieving greater efficiency and productivity, the Federal Government Corporation structure can be used to realize a renewed focus on core responsibilities for the Army as well as the other services. Just as GM used its acquisition and divestment of EDS as an expression of corporate strategy, so too can the Army use the FGC as an instrument of national military strategy.

FEDERAL GOVERNMENT CORPORATIONS

The basis for Congress's ability to create government corporations is derived from the Necessary and Proper clause of the Constitution, Article I, Section 8, Paragraph 18, which states:

> To make all laws which shall be necessary and proper for carrying into execution the foregoing powers and all other powers vested by this Constitution in the Government of the United States or in any Department or Officer thereof.

There is a long history of Supreme Court rulings and case law using this paragraph as the foundation of the ability of Congress to create corporations.[1]

There are three basic groups of organizations that are considered FGCs. The first group is the Government Sponsored Enterprises (GSEs), which are very large financial organizations such as Federal Home Loan Banks, Fannie Mae, Freddie Mac, FICO, REFCORP, and six other specialized lending organizations. These organizations have special financial privileges and were created to facilitate the creation of credit for specific economic groups or for a specific financial purpose such as recapitalizing insolvent savings and loans. Congress usually categorizes GSEs as mixed-ownership FGCs. In reality the amount of private ownership varies from 0 to 100 percent. The second group has only one member, the United States Postal Service (USPS). This organization is officially categorized as an Independent Establishment of the Executive Branch of the U.S. Government. The special category is drawn in part from the specific constitutional citation which empowers the Congress to create "Post Offices and post roads."[2] In the same way, the constitutional provision for Army arsenals ("the erection of forts, magazines, arsenals, dock-yards and other needful buildings.") could form the basis for a new Independent Establishment.[3] The third group comprises about 50 government corporations that are chartered by Congress to

---

[1] See, for example, *Osborn* v. *Bank of the United States*, 22 U.S. (9 Wheat.) 738, 859-60 (1824); *Federal Land Bank* v. *Bismark Lumber*, 314 U.S. 95, 102-103 (1941); and *Pittman* v. *Home Owners' Loan Corp.*, 308 U.S. 21, 33 (1939).

[2] United States Constitution, Article 1, Section 8, Paragraph 6.

[3] United States Constitution, Article 1, Section 8, Paragraph 17.

achieve specific national policy goals. For example, in the first Clinton Administration, when it was felt that a domestic "Peace Corps" might solve some of the problems of the inner city, the Congress at the behest of the administration created the Corporation for National and Community Service (AmeriCorps) in 1993.[4] The most recent FGC is the Valles Caldera National Preserve and Trust, which authorizes the acquisition and independent management of the Valles Grande, an enormously beautiful and undeveloped area of land in northwestern New Mexico.[5] These organizations include such familiar entities as the Tennessee Valley Authority (TVA), the National Railroad Passenger Corporation (AMTRAK), and the Smithsonian Institution. Others may not be so familiar, such as Federal Prison Industries, Inc., the Saint Lawrence Seaway Development Corporation, and the Pennsylvania Avenue Development Corporation. In this list we do not include the national banks that have a federal charter but no government-appointed board of directors members or ownership, or the more than 80 patriotic or charitable organizations that have a federal charter but receive no federal funds and are responsible for their own business affairs.[6,7] In addition, the federal government has directed the establishment of some corporations not directly chartered by Congress but nevertheless owned, funded, or directed by the government, such as the Corporation for Public Broadcasting and the American Institute in Taiwan.[8,9] We do not consider these in our analysis of the FGC option for the Army, but we observe that the government corporation form can assume a wide variety of identities.

Since World War II, Federal Government Corporations have been used as instruments of national policy because of their efficiencies arising from commercial market forces, their flexibilities with regard to encumbering regulations, and their ability to access financial alternatives. The usual process for creating an FGC starts with

---

[4]National and Community Service Trust Act of 1993, P.L. No. 103-82.
[5]The Valles Caldera Preservation Act of 2000, P.L. No. 106-248.
[6]12 U.S.C. 21, 35, 40, 41, 215c, 1817, 3102.
[7]36 U.S.C. 1–9, 13.
[8]47 U.S.C. 396–399.
[9]22 U.S.C. 3301–3310.

Congress drafting a charter that sets forth the entity's goals, obligations, special powers and exemptions, and organizational structure including the composition of the board of directors. The enabling legislation can specify a federal charter or incorporation under the laws of the District of Columbia. All FGCs are created as separate and permanent legal entities. Generally, in the congressional charter the right to sue and to be sued is a provision and is considered a waiver of sovereign immunity that clearly distinguishes the FGC from other government organizations.[10] Additionally, the FGC can settle cases against it independent of Department of Justice representation.[11] These rights and privileges obtain in regard to issues associated with the conduct of normal business. For issues regarding constitutional rights the courts consider an FGC a state actor.[12]

Efficiency. Free market forces generally create low-cost products and services. As echoed in the Truman quote that opened this chapter, when products and services provided by the government as part of a national policy goal are inherently commercial in nature, the option of choice is most likely the government corporation. Because of these efficiencies, the FGC option appeals to a broad base of support. Fiscal conservatives can agree that the low-cost option for a national policy goal can be created using free-market forces. Adherents of small government can agree that the FGC could be a first step to the privatization of commercial government activities. Even democratic socialists can see the FGC as a method to redistribute the wealth created by public works activities or those arising from natural monopolies. Beyond the economic efficiencies, the FGC option creates a highly focused organization with a well-defined national policy goal. FGCs are allowed to focus on a single product or service and on a limited customer base or constituency. This is in contrast to the usual multimission span of a traditional government agency.

Flexibility. FGCs are granted much flexibility with regard to the otherwise encumbering regulations that would obtain for a traditional

---

[10] *United States* v. *Nordic Village,* 503 U.S. 30, 34 (1992); *Federal Housing Administration* v. *Burr,* 309 U.S. 242, 245 (1940).

[11] 28 U.S.C. 516 (1993); 5 U.S.C. 3106 (1988).

[12] *Lebron* v. *National R.R. Passenger Service Corp.,* 115 S. Ct. 961 (1995).

government agency. FGCs can enter contracts for goods and services independent of the FAR. They can buy and sell assets independent of the Federal Property and Administrative Services Act of 1949.[13] Most FGCs are exempted from Civil Service regulations on pay and employee tenure (Lilienthal and Marquis, 1941) and from government personnel ceilings. Some FGCs are even exempted from the Government Corporation Control Act (GCCA), which was created to better regulate the mix of powers and privileges granted to FGCs in their congressional charters.[14]

Finance. FGCs benefit from financial freedoms beyond the restrictions on federal agencies. FGCs have the right to borrow funds from commercial and private sources, to issue debt in the form of bonds, and to own, to acquire, and to dispose of real property plant and equipment.[15] Generally an FGC is not subject to the year-end budget pressures forcing expenditures within a given fiscal year. They can enter into multiyear commitments based on funding that will be available in their budgets regardless of yearly expenditures. Mixed and private ownership FGCs are usually financed "off the balance sheet" (Collender, 1997) which, in effect, excludes them from the national accounts. With such a status, the debts of such organizations do not count against the national debt and are not subject to deficit reduction goals or spending caps when Congress is operating under budget reduction measures such as the Gramm-Rudman-Hollings budget reduction process. Some FGCs are exempted from local, state, and federal taxes, and their executives are excluded from Security and Exchange Commission regulations.

Federal Government Corporations can be analyzed along three basic dimensions: control, cash, and customers. FGCs are categorized for legal and regulatory purposes as government-owned, mixed-ownership, and private-ownership (U.S. GAO, 1995). However, these categories are not useful in determining, for example, how to deal with an FGC as a customer or how to think about FGCs as an instrument of strategic policy for an organization.

---

[13]Federal Property and Administrative Services Act of 1949 (41 U.S.C. 251–260).
[14]Government Corporation Control Act (GCCA), 31 U.S.C. 9101–10 (1988 and Supp. Vol. 1993).
[15]Ibid.

Spinning Off Army Activities into Federal Government Corporations   61

The strategic control of an FGC flows from the level of ownership by the federal government, the level of ownership by private parties, and by the composition of the board of directors (BOD). Operationally, the control of the FGC is in the hands of the leadership brought in to run it. These individuals report to the BOD. For a government-owned FGC, the President of the United States appoints the majority or the entire BOD, whereas for a privately owned FGC, the President appoints a minority of BOD positions. The mixed-ownership FGCs are in the middle. In the control dimension, FGCs are spread from mixed control to total private control, as illustrated in Figure 4.1.

The locations along the control dimension for a government department or agency, a GOCO, an FFRDC, and a GSE are displayed for comparison in the figure. The department or agency is totally under government control with both line and program management reporting directly to government officials. The GOCO is similar, although the distance from total government control is increased

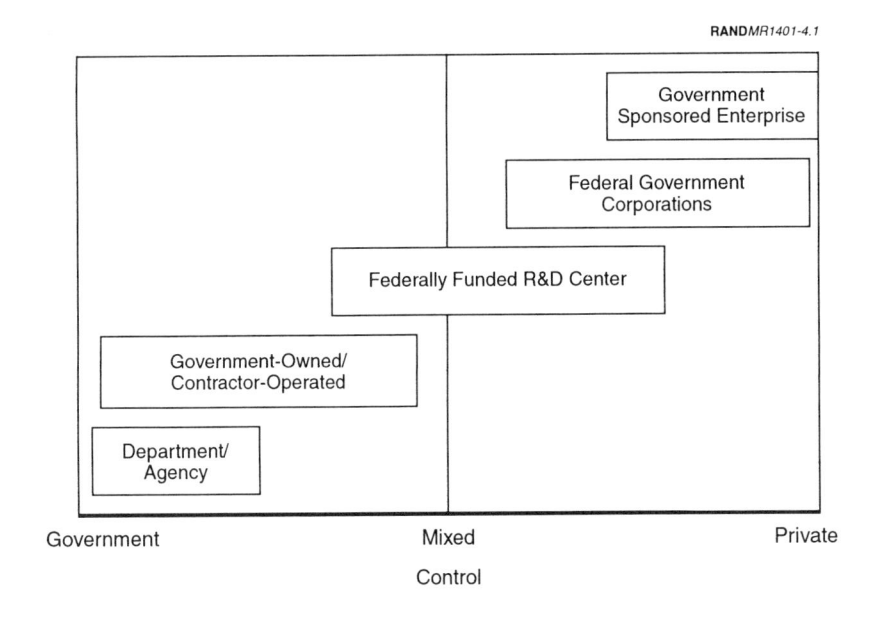

Figure 4.1—Comparison of Organizations Along the Control Dimension

because a contractor now operates a government-owned facility. An FFRDC is similarly more distant from complete government control. An FFRDC is created to give an exclusively government customer an unbiased research opinion on critical issues. FGCs are spread from mixed to totally private whereas GSEs are almost entirely under private control.

Along the cash dimension, the basic organizational characteristic is source of revenue. Figure 4.2 shows where the FGC and other organizations fall in this dimension. The figure illustrates that FGCs span the spectrum from mostly governmentally funded to mostly privately funded. Generally, FGCs that are mostly government funded tend to be mostly government controlled, such as Federal Prison Industries, Inc., or the Saint Lawrence Seaway Development Corporation, whereas the opposite is true for FGCs funded mostly from private sources, like the Tennessee Valley Authority and AMTRAK. In contrast a department or agency is totally funded from government sources. This is true also for a GOCO and for an FFRDC.

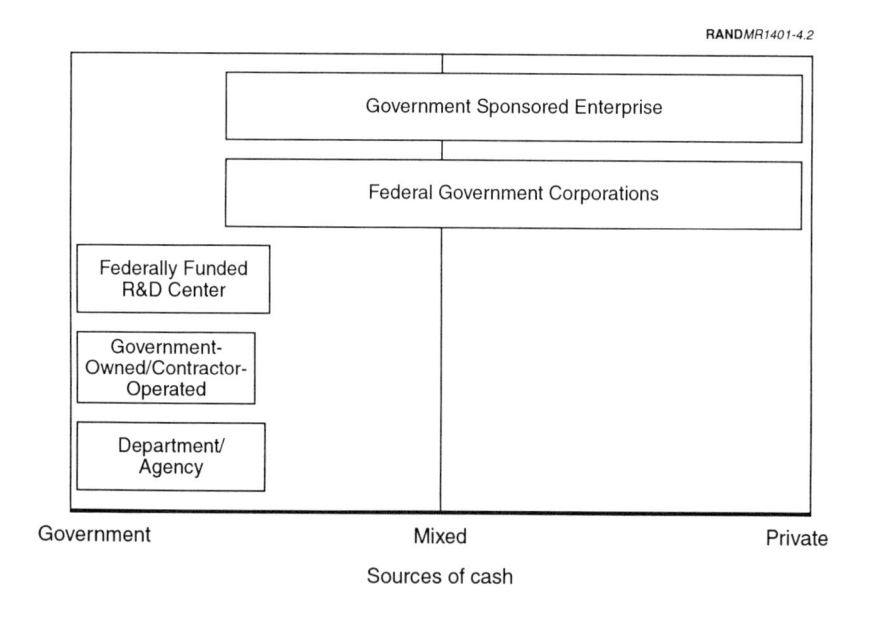

Figure 4.2—Comparison of Organizations Along the Cash Dimension

Along the dimensions of control and cash, FGCs are roughly in the mixed category, although some are on either end of the spectrum. As the Truman quote that opened this chapter suggested, the important distinction of FGCs is that they have or could potentially have a customer base that is mixed or mostly from the private sector, as indicated in Figure 4.3. GSEs have a customer base that is entirely in the private sector, whereas GOCOs and FFRDCs have only government customers.

FGC customers are almost always the commercial sector or the general public. Some FGCs have government customers as well. The basic theme for all FGCs is that corporations can be more efficient than governmental structures when it comes to market transactions. If this product or service can be offered to the government as well as the private sector, so much the better. Under those circumstances the government can extract for its own use the efficiencies embodied in the product or service arising from the commercial market pressures.

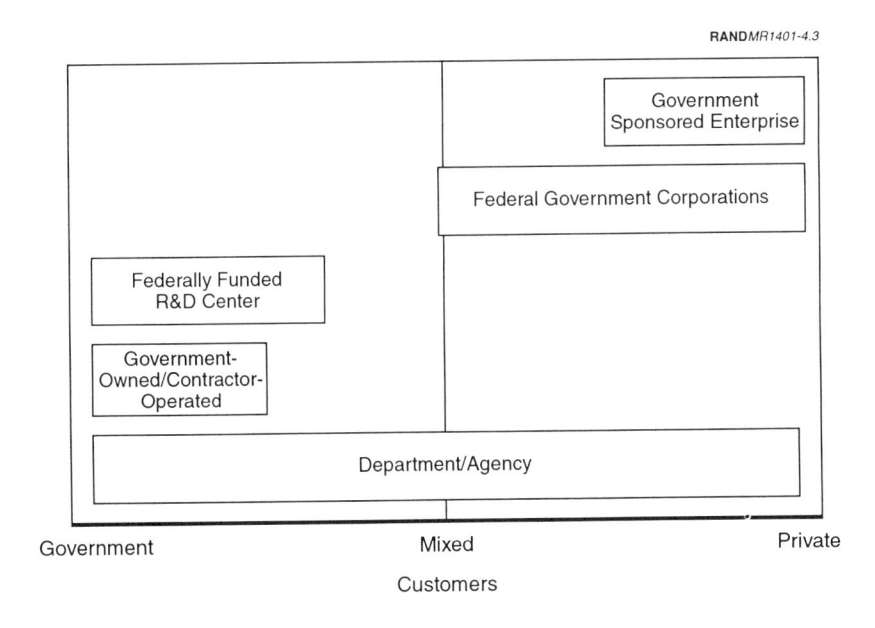

Figure 4.3—Comparison of Organizations Along the Customer Dimension

Whereas FGCs have existed for more than 200 years, there are significant differences in how they are structured and controlled. There is essentially no uniform legal definition of an FGC. Because Congress individually charters each FGC, the range of applicable statutes may vary widely. In 1945, the Congress established the Government Corporation Control Act, which tried to better define FGCs in terms of corporations either owned or controlled by the U.S. government. However, specifics have not been standardized, and many FGCs are exempted from the GCCA (Moe, 1983; U.S. GAO, 1983).

THE ARMY AND FGCs

The FGC presents the Army with a very flexible instrument to implement policy and strategic initiatives. For example, it has long been an Army policy that core logistics capabilities will be sustained in peacetime so that they can be available in times of war. This policy has resulted in a general understanding of what is core in the Army and what is peripheral to the core. For many of the peripheral functions, outsourcing of certain services and products has proved to be a good mechanism for increasing efficiency and reducing costs. For other activities that are on the boundary of these two domains, simple outsourcing is far too trivial a solution. For example, the Army depots and arsenals have many Army-unique capabilities that are significantly underutilized in peacetime yet may be needed in future times of war. If these organizations were converted to FGCs, then their underutilized workforces and physical plant could be applied to creating economic value in the private sector. Partnering with industry could take the form of a strategic partnership with a company that otherwise would be the FGC's competitor, or a mixed-ownership FGC with some BOD members drawn from the industrial stockholders. As an FGC matures and our understanding of the warfare of the future evolves, some of the capabilities of yesterday may migrate from the core to the peripheral domain. Then the FGC can outsource this capability or divest it. As new needed capabilities become apparent, they can be acquired from the private sector. These divestments and acquisitions can be done in a flexible and expeditious manner because, like most corporations, the FGC can maximize best long-term value rather than being driven to lowest cost.

The FGC option can provide the Army leadership with a flexible and agile instrument for policy initiatives. These organizations are controlled by their congressional charter and by their board membership. Thus, the Army leadership can extract itself from day-to-day operations and assume a more strategic perspective.

FGC AREAS OF CONCERN

A central premise in our constitutional form of government is that organizations that implement public policy should be held accountable for their actions. Moreover, public organizations supported by public funds should not benefit private organizations. All benefits from public funds should flow to the public. The FGC sits atop this divide between federal and private roles and responsibilities. Let us suppose that an FGC called United States Ordinance Corporation (USOC) is created from the arsenals, depots, and ammunition plants. Consider the case of a machinist at USOC who posts a notice about a meeting for a political action group on the company bulletin board. The vice president for human resources has the notice removed and admonishes the employee. The employee insists that it be posted as a matter of First Amendment rights. USOC has a policy on posting notices, allowing the vice president to decide. Is USOC acting as a part of the federal government that must be bound by the Constitution, or is it acting as a private company within its rights? Consider the case of a commercial client who sues USOC for nonperformance on a contract. Can USOC claim sovereign immunity and escape any legal remedy? If USOC makes an enormous profit one year, should those profits be returned to the U.S. Treasury? General Electric proposes a strategic alliance with USOC for heavy industrial machining using existing and new USOC staff. In return for stock and a board seat, General Electric will build two new facilities, populate them with the most advanced equipment, and provide the workforce with the needed training. How should USOC respond? Just as the economic efficiencies of the free market can create low-cost products and services, so too do these economic forces drive corporations to maximize profits. When profit maximization is at odds with a part of the formative national policy, where should USOC's loyalties lie? These questions are but a few of the manifold of possible issues that can and will arise in the life of USOC.

To be prepared for these possibilities with a clear path of action, USOC needs a well-crafted congressional charter making clear the roles and responsibilities of the corporation itself, the executive management, and the board of directors. However, Congress has not crafted a clear FGC charter for many decades, despite creating about one FGC per year since World War II. Although these FGCs have served the government well as instruments of federal policy, this service has been executed with some difficulty. Many of the difficulties derive from an unclear path of accountability, to the President, to the Congress, and to the American people.

Crafting good FGC charters has been the subject of considerable effort in the public administration arena (Froomkin, 1995; U.S. GAO, 1983; Leazes, 1987). Several sample charters along with the examples of the types of considerations that should be raised in drafting these documents are available (NAPA, 1981). All issues and concerns can be addressed in a well-crafted corporate charter and a well-designed BOD. The lessons learned from the present set of FGCs can provide considerable insights into the proper course. Clarity is the key to a successful Federal Government Corporation.

POTENTIAL ARMY CANDIDATES FOR FGCs

At least three Army candidates for FGCs have been proposed: (1) Army chemical demilitarization, (2) Army R&D laboratories, and (3) Army depots. As part of the 1998 AMC Redesign OIPT, the Army considered turning the Army chemical demilitarization operations into an FGC. While the assessment was positive, no action was subsequently taken (Gonczy, 1998). At this stage in the demilitarization process, it may now be too late to consider making this organizational change. However, the other two FGC candidates are still timely and relevant, and the Army has not seriously studied them. In this section, we indicate that both should be considered as possible candidates for establishing FGCs.

Army R&D Laboratories

Context. The Army R&D laboratories include AMC laboratories, the Corps of Engineers (COE) laboratories, and the Army medical laboratories. The AMC laboratories include both the Army Research Laboratory (ARL) and the Research, Development, and Engineering Centers (RDECs) associated with the various AMC commodity commands. In this discussion, we will limit our comments to the AMC laboratories, although the observations most likely apply to the COE and medical laboratories as well.

Over the past few years, the AMC laboratories have been criticized for not initiating and broadening exchanges with industry (NRC, 1998), for not working more closely with the RDECs in avoiding duplication and competition for research funds (NRC, 1998), for working too much in isolation from other research laboratories (Crawford, 1998), and for not attracting competent engineers and scientists to DoD laboratories and S&T management (DSB, 1998). While much of the criticism has been directed at the ARL, the RDECs (as well as other DoD laboratories) have not gone unscathed. While many of these concerns have been addressed by the AMC, the laboratories still remain under the microscope of various government and industry critics.

Analysis approach. At the Army's request, the Arroyo Center has looked into the pros and cons of candidate alternative organizational models for the AMC laboratories.[16] The broad aim of the studies was to provide the Army with an independent analysis to help guide the long-term evolution of the laboratories. In each case, hypothetical candidate models were assessed in terms of evaluation criteria that attempt to represent the generic functions of the laboratories. The high-level generic functions were synthesized from available information on laboratory functions.

The alternative laboratory organizational models that have been considered are shown in Table 4.1, along with the defining feature of each concept. The models are listed in alphabetical order. A more complete description of each is given in Appendix A.

---

[16]Over the past few years, RAND has performed several studies that have investigated laboratory alternatives. Because of the potential sensitivities involved, the findings have not been released as public-domain documents, and are only discussed in general terms here.

Table 4.1

Organization Laboratory Models Considered

Organizational Models	Defining Feature
Defense Research Institute	• Is like a graduate school doing hands-on R&D
Defense Working Capital Fund	• Only does work that is paid for
Federal Government Corporation	• Is owned/controlled by the public sector
Federally Funded Research and Development Center	• Is modeled after MIT Lincoln Laboratory
Government-Owned/ Contractor-Operated	• Is modeled after Sandia/LLNL/Los Alamos
Government as a Subcontractor	• Has labs compete with private industry
International Laboratory	• Performs R&D of mutual interest
Joint Service Laboratory	• Jointly works with combined funds
Outsource Laboratory	• Is modeled like DARPA, with 95 percent outsource
Private Laboratory	• Has no ownership stake or control
Reserve S&E Corps	• Has S&Es on call for service
Technology Incubator	• Provides basic service support
Venture Capital	• Operates like commercial VC fund
Virtual Laboratory	• Performs R&D at multi-lab sites

The various candidates were assessed in terms of the following criteria. Can they

- Understand and influence the Army's long-term visions to maintain technological superiority?
- Plan and direct a research program to implement the Army vision?
- Influence and leverage commercial technology/system developments?
- Conduct high-quality, revolutionary, government-funded research?
- Perform the "smart buyer" function for current and future materiel acquisitions?

- Plan and direct the integration of technologies into current and future weapon systems?
- Evolve as necessary to effectively and efficiently achieve mission goals?

The models were qualitatively assessed against the assessment criteria by RAND staff members who possessed some general familiarity with military, civil, and civilian R&D laboratories as a result of their past work experience in military, FFRDC, and industrial R&D centers.

Analysis results. Based on the subjective assessment performed by the RAND staff members, the FGC model emerged as one of the more promising organizational models. Its strength lies in its ability to achieve flexibility and efficiency, characteristics desirable to the Army in adapting to changing research needs.

Clearly, the Army needs to look more closely at the FGC model before considering it as a viable option. Other factors must be considered, including ease of implementation, cost of implementation, approval authorization, public support, etc. However, it is not insignificant that the FGC looked so promising in the preliminary exercise.

Army Depots

Context. We will use the term "Army depot system" to refer to the Depot Maintenance—Army (DMA) activities in the AMC, whose operating expenses (i.e., wages, salaries, and benefits; materiel costs; and all other operating expenses) are financed and paid for by the Army Working Capital Fund (AWCF).

In physical terms, the Army depot system consists of the following five heavy maintenance depots:

- Anniston Army Depot, Anniston, Alabama;
- Corpus Christi Army Depot, Corpus Christi, Texas;
- Red River Army Depot, Texarkana, Texas;
- Letterkenny Army Depot, Chambersburg, Pennsylvania;
- Tobyhanna Army Depot, Tobyhanna, Pennsylvania.

These five depots, with a civilian workforce of more than 9,500 people, are in the business of repairing, overhauling, and upgrading weapon systems and equipment. They do this mainly for the Army, but they have other customers as well. In addition to their maintenance work, the depots provide tenant support to other AMC, Army, and DoD activities at the five locations. In financial terms, the system expected to collect about $1.2 billion in revenues from its customers in FY00.[17]

The Army depot system is an overwhelmingly civilian activity. The Army's FY01 President's budget submission indicates it will have 9,502 civilians on the payroll in FY00, but only 21 military personnel. That makes the Army depot system not only the largest civilian employer in the AWCF (accounting for just over half the total number of civilians in the four AWCF business areas of depot maintenance, supply, ordnance, and information services), but also the largest single employer of civilians in AMC. (After the Army depot system, the next-largest civilian employers at AMC, which employ roughly 50,000 civilians in total, are the commodity commands Tank-automotive and Armaments Command (TACOM), Communications-Electronics Command (CECOM), and Aviation and Missile Command (AMCOM), with 7,000 to 8,000 civilians each.) Overall, about 40 percent of the total civilian positions at AMC are funded by the AWCF, about 20 percent are funded by other, smaller reimbursable-type activities that the command performs, and the remaining 37 percent are funded by direct Army appropriations allocated to AMC.

Like all Working Capital Fund (WCF) activities, the Army depot system is already required (by DoD and Army policy) to operate in many respects "like a business." In particular, the system relies on customers to come in the door with work and the money to pay for it. However, Army customers are not required to buy from the Army depot system. In fact, many routinely use alternative providers to get depot-level work done. WCF policies also establish a financial "bottom line" under which the Army depot system must try to keep its share of the AWCF in balance by seeking to achieve an "Accumulated Operating Result" (AOR) of zero over time. Most

---

[17] Based on the Fund 14 "Revenue and Expenses" exhibit for Depot Maintenance in the FY01 AWCF President's Budget (February 2000), p. 39.

important, given the Army's military mission, the Army depot system has a key responsibility to do its part in making sure Army customers have the quality products and services they need to do their jobs.

Unlike a normal business, however, the Army depot system (like all DoD WCF activities) must operate in the "planned economy" of the DoD, in which the annual programming, budgeting, and appropriations process places highly non-market-like constraints on such things as the total amounts available to be spent, where investments can and cannot be made, and the prices that can be charged for goods and services provided. As a result of these and other special political and legal constraints on what managers inside the depot system can do, operating-cost reductions in the Army depot system have not kept pace with reductions in force structure and workload.[18]

Analysis approach. In a way similar to how we looked at organizational alternatives for Army laboratories, we have assessed alternative candidates for the Army depot system against assessment criteria using an in-house RAND evaluation team. In this case, however, the assessment was not performed as in the laboratory evaluation case. Rather than using a formal Delphi approach with four rounds of evaluations, we used a traditional consensus-forming approach with the evaluators ranking the various alternatives after discussing them in an open forum.

The alternative depot organizational models that we considered are shown in alphabetical order in Table 4.2, along with their defining feature. The "Baseline Plus" model refers to a restructured and improved depot system operating under full organic control by the Army. This model is assumed to be redesigned according to current and evolving Army plans to pursue optimal efficiency and capacity consistent with maintaining an in-house government system. Each organizational model is described in more detail in Appendix B. The FGC model has a form of ownership different from the other models.

---

[18]For example, W. N. Washington in the *Acquisition Review Quarterly* (Summer 1999) cites data provided by the AMC Deputy Chief of Staff for Resource Management (B. Leiby, November 30, 1998) indicating that from FY92 to FY98 operating costs in the Army depot system fell by just under 30 percent, while incoming workload fell by almost 50 percent.

Table 4.2
Organizational Depot Models Considered

Organization Model	Defining Feature
Baseline Plus	Achieves optimal efficiency/capacity within constraints of current system
FGC	Has flexibility in control, financial operations, and customers
GOCO	Has depot facilities/equipment owned by the government but has depot operated by contractors with mostly contractor staffing
Prime Vendor Support	Performs all wholesale logistics functions under a system contract
Privatization in Place	Has a private contractor assume ownership of a depot at its current location
Public-Private Partnership	Has private sector contribute property, plant, equipment to achieve end goal of the public-private partnership

It is chartered by Congress, which sets forth goals, obligations, special powers, exemptions, and composition of the board of directors. As mentioned earlier, the FGC benefits from financial freedom beyond the restrictions on federal agencies. It also offers workforce management options unavailable to government agencies. The unique characteristics of FGCs make this approach a promising candidate for application in the Army depot system.

To test that hypothesis, we used various assessment criteria in the evaluation, which are shown in Table 4.3 listed under five generic categories covering mission fulfillment, institutional issues, statutory requirements, financial issues, and congressional support.

Analysis results. The results of the assessment indicate that the FGC, as a candidate approach dealing with issues in the current Army depot system, ranked with the highest among all the approaches.[19]

---

[19]Michael Hynes, "Organizational Alternatives for the Army Depot System," private communications, May 2000.

Table 4.3
Depot Assessment Criteria

Generic Category	Assessment Criterion
Mission fulfillment	• Improves repair/maintenance capability • Improves system renewal (recapitalization/modernization) • Improves surge/reconstitution capability
Institutional issues	• Improves workforce management • Improves process management • Offers entrepreneurial (agility to respond to business opportunities)
Financial issues	• Creates competitive environment • Reduces costs • Provides access to capital to finance expansion/innovation
Congressional support	• Lessens concern over jobs • Improves local economic help • Addresses core competency concerns
Statutory requirements	• Satisfies requirement for competition (10 U.S.C. 2469) • Satisfies limitations on contracting (10 U.S.C. 2466) • Satisfies core logistics capability (10 U.S.C. 2464)

Other models also achieved high rankings, in particular, Baseline Plus and Public-Private Partnership. As in the case of the emergence of the FGC as a candidate for the AMC laboratory organization, the Army needs to look more closely at the FGC model and address a series of significant issues.

Two aspects of the Army depot system explain why caution is necessary before applying the FGC idea. First, the FGC concept works best when external commercial opportunities exist, and it is not clear which, if any, of the five Army depots meet that criterion. Second, the Army depot system relies upon Program Offices and Item Managers in the AMC Major Subordinate Commands (MSCs) for guidance, direction, and workload assignments. Thus, any FGC arrangements would necessarily entail establishing special working arrangements and relationships with the MSCs. With those caveats, we believe that the methodology we have developed is a useful way for the Army to further evaluate the FGC approach.

Notwithstanding the cautionary notes above, there are three additional reasons why the FGC model is appealing for Army depot maintenance. The first reason has to do with the uncertain nature of the depot maintenance business itself. The DoD's traditional Planning, Programming, Budgeting, and Execution System (PPBES), under which the Army and all the services must operate, works best when "requirements" can be clearly defined, programs and workloads can be established with certainty ahead of time, and management actions can be taken to control how money is spent in execution. In the depot maintenance business, however, it is virtually impossible to predict demands, workloads, and sales with the degree of accuracy required to ensure "break-even" performance.

The Army depot system's financial performance provides evidence that this is true. In FY98, for example, the system reported losses of $133.7 million on revenues of roughly $1.5 billion.[20] The problem is very much entwined with the PPBES system and the way DoD must operate as part of the executive branch within the government. In particular, because authority to spend AWCF dollars is controlled for the DMA system in essentially the same way it is for appropriated dollars, the Army depot system (despite the fact that it is a WCF activity) does not really enjoy any more "business flexibility" than it would if it were simply an appropriated function.[21] Therefore, a key

---

[20]AWCF FY00/01 Budget Estimation Submission, September 1999, reports the FY98 DMA Net Operating Result (NOR) was negative: $133.7 million. The "recoverable" portion of the Accumulated Operating Result (reflecting the net effect of annual NORs going back to FY92, adjusted and reduced by write-offs) was reported to be negative $36.1 million.

[21]As an example of this, the HQDA Program Analysis and Evaluation Directorate stated to the DWCF Reform Task Force in September 1998: "Frustrations with the current structure boil down to an inability to accurately forecast customer demands and restrictions against sizing workforce and facilities appropriately. Imbalances cause cash problems or operating losses which manifest themselves as bills to the services." Although it is true that depot operations have been "industrially funded" for many years, it was only in FY92, with the creation of the Defense Business Operations Fund and its establishment of "full cost recovery" pricing combined with the "stock funding" of DLRs in the Army, that these problems began to surface.

structural reason for considering FGC-type arrangements is to remove the activity in question from the rigidity of the annual budgeting and appropriations process, when that rigidity conflicts with the basic nature of the business.

The second reason for considering FGC approaches for the Army depot system has to do with the mandate facing the Army from the 1997 Quadrennial Defense Review (QDR) to eliminate 17,366 civilian positions by FY04, 8,530 of which are supposed to come from AMC. As the largest single employer of civilians in AMC, the Army depot system is a natural candidate to be a significant contributor to the AMC effort, but, in fact, it is not. Army and AMC plans as of 1999[22] indicate that only 400 of the total 8,530 positions to be eliminated from AMC will come from the Army depot system, and those reductions are attributed entirely to the decentralization and transfer of depot management from the Industrial Operations Command (IOC) to the individual commodity commands. By way of comparison, 1,567 positions are to come by restructuring AMC RDECs (which employ far fewer civilians than the depot system), and another 730 positions are to come by competing functions and reducing test requirements in the Test and Evaluation Command (TECOM), which employs less than half as many civilians as the depot system.[23]

By applying the FGC concept to its depot system, the Army could make reductions to its *government* civilian workforce, without having to eliminate jobs. Indeed, one of the significant advantages of the FGC approach, if applied at a particular depot, is that it would allow the employees at that depot to keep their jobs at that location provided they would be willing to give up their status as *government* employees. To be sure, a key advantage of the FGC approach is that it affords much greater latitude and incentives for managers and employees to reform internal processes, and that almost always entails changing some jobs and eliminating others. However, it also

---

[22]As described in the March 1999 GAO Report, *Status of Efforts to Implement Personnel Reductions in the Army Materiel Command*.

[23]Salaries, wages, and benefits for civilian personnel are the largest single element of expense in the Army depot system. In FY00, salaries, wages, and benefits for the system's 10,409 civilians will be $534.5 million (an average of $51,350 per person), with the next-largest element of expense being the cost of materials and supplies used in repair operations—$404.1 million in FY00.

means that new jobs could be created if the new FGC takes advantage of the opportunities to seek new customers and markets, compete, and grow. Because FGCs can compete for new work, partner with industry,[24] and reward employees in ways that internal government activities cannot, there is every reason to believe that depot employees who are willing to commit to the idea have more to gain from being in an FGC than they have to lose.

The third reason for looking at the FGC idea for the Army depot system is that senior Army leadership has already considered the concept and indicated its willingness to pursue it further—an important prerequisite for possible success. In particular, a consensus has been reached in the senior programming, budgeting, and logistics communities at HQDA that the FGC concept is not an unreasonable one to explore for the AWCF activities. Indeed, that consensus has existed for some time. In September 1998, as part of its contribution to a joint DWCF Reform Task Force under the Defense Reform Initiative, the Army itself proposed that the Task Force consider (as one possible reform alternative) converting some or all the DoD's DWCF activities to FGCs.[25]

The upshot of the Army proposal, after a year of Task Force study that included a high-level review by the Task Force's Executive Steering Group with the Assistant Vice Chief of Staff representing the Army, is that the Task Force and its Steering Group recommended to the Deputy Secretary of Defense and the Defense Management Council (DMC) that they:

---

[24]"Strategic partnering" with industry and other options for "commercializing" the Army depot system are beginning to receive increased attention inside the Army as a way to help reform and improve the system. See, for example, Washington (1999).

[25]In a presentation to the DWCF Reform Task Force, September 3, 1998, prepared by the Program Analysis and Evaluation Directorate (PAED), Office of the Director of the Army Staff, the Army told the Task Force: "Prior attempts to address infrastructure limitations have focused on complete commercialization or privatization of DWCF activities. This proposal offers a way to maintain control over inherently governmental functions (i.e., mobilization capability) while improving ability to operate 'like a business.'"

Support proposals through the DMC for alternative organizational structures [for DWCF activities], such as a Federal Government Corporation.[26]

To ensure that the Army and other services were still on board with the FGC idea before issuing the above recommendation, the Task Force leadership individually briefed flag-level representatives from the programming, budgeting, and logistics communities in the Army and other services in June 1999 on all the Task Force recommendations.[27] The HQDA representatives from all three communities gave a "green light" to the Task Force recommendation that a "detailed feasibility analysis" should be done on adopting an FGC structure for a "pilot" WCF activity, thus indicating their concurrence with the Task Force statement that

> Of the alternatives (status quo, Federal Government Corporation, privatization, employee stock ownership, mutual benefit corporation performance-based organization), the Federal Government Corporation provides the most gains in operational and financial flexibility while continuing to address the Department issues of industrial preparedness and mobilization.[28]

---

[26]DWCF Reform Task Force Decision Briefing (showing decisions by the ESG made on August 18, 1999).

[27]DWCF Reform Task Force Information Briefing to the DWCF Policy Board and the Deputy Secretary of Defense, July 1999.

[28]By June 1999, the Task Force had in hand the results of a study it had commissioned on the FGC concept. See Vivar and Reay (1999).

Chapter Five
CONCLUSIONS AND RECOMMENDATIONS

> But the only way of discovering the limits of the possible is to venture a little way past them into the impossible.
>
> — Arthur C. Clarke
> *Profiles of the Future*
> 1972

Three collaborating/partnering concepts appear promising for possible Army exploitation: (1) forming real-estate partnerships using public-private partnerships, (2) using venture capital approaches to capture high-tech commercial technologies and products, and (3) forming spin-offs using FGCs.

Each concept requires resolving key issues before the Army for implementation can seriously consider it. In the case of public-private partnerships, various implementation issues must be resolved within the Army, including whether financially sound concepts can be proposed by the installations. In the case of the venture capital concept, the potential merits of such a concept to meet the Army's technology needs must be addressed in further detail. Monitoring the status of the CIA's venture capital efforts will help in this assessment. In the case of FGCs, the value of establishing the Army laboratories and depots as FGCs will depend on how many external commercial opportunities exist and further analysis on how to best structure continuing relationships with other Army organizations.

Once these key issues are satisfactorily addressed, the Army should create pilot programs to test the concepts. This approach is consistent with the new industry paradigm that argues that one learns more about something by acting on it (in this case, by establishing pilot programs) instead of, as in the past, waiting until it is thoroughly understood before acting.[1] This paradigm argues that one learns more and faster from tinkering and experimenting. For fast-moving technologies, not acting is equivalent to staying behind.

Our recommended next steps are as follows:

- Test the hypothesis that there are a lot of good ideas for public-private partnerships at Army installations.
- Take steps to establish an Army Venture Capital Fund by:
  — Identifying technologies and problems amenable to solution through Army venture capital funding,
  — Establishing an Army venture capital advisory committee, and
  — Initiating the establishment of the AIIC by either
    » Drafting appropriate legislation and working with Congress to enact it as a part of the next DoD budget, or
    » Identifying current funds and, using an Other Transaction, contract with an appropriate organization (either an FFRDC or an existing venture capital firm).
- Review the results of the assessment methodology of alternative laboratory and depot models and assess the operational payoff of establishing laboratories and depots as FGCs.

---

[1] In the foreword to Michael Schrage's 1999 book on how the world's best companies simulate to innovate, Tom Peters refers to the act/learn paradigm in terms of "Ready.Aim.Fire!" The act/learn paradigm is based on the scientific method of experimentation. After 100 years of plodding along with emphasis on analysis, or the "Aim" in "Ready.Aim.Fire!," the business schools are now extolling the virtues of experimentation as a way of testing new concepts ("Ready.Fire!Aim").

Appendix A
DESCRIPTION OF LABORATORY MODELS CONSIDERED

In this appendix, we describe in more detail the laboratory models used in the analysis described in Chapter Four.

DEFENSE WORKING CAPITAL FUND (DWCF) MODEL

The DWCF model is based on agencies such as the Department of Defense (DoD) depot maintenance program and the Naval Research Laboratory (NRL), whereby the agencies only do work that is requested and paid for by their customers.

DEFENSE RESEARCH INSTITUTE (DRI) MODEL

Conceptually, a DRI is a world-class graduate school that specializes in fields of research important to the Army. Admission is competitive and similar to today's state and private universities. The graduate students would engage in hands-on research and have thesis and dissertation goals similar to those at most graduate schools. In this scheme, the Army would be able to attract and train highly qualified individuals and offer permanent employment to the most talented. Tenure could be granted to exceptional faculty members, which could facilitate the maintenance of corporate memory.

FEDERAL GOVERNMENT CORPORATION (FGC)

The FGC model is discussed in the main body of the report. The key feature of an FGC is flexibility. FGCs are granted flexibility with regard to otherwise encumbering regulations and with regard to Civil Service rules and federal acquisition and disposal requirements. They are granted freedom from the political forces driving congressional actions. They are allowed to focus on a single product or service and on a limited customer base or constituency by being insulated from the demands of a multimission agency. Finally, FGCs are allowed financial freedoms unavailable to federal agencies. In particular, they can borrow money from commercial sources, they can issue debt in the form of bonds, they can be exempt from local, state, and federal taxes, and they can benefit from "off-the-balance-sheet" status, multiyear federal funding, and exemption from deficit-reduction spending caps.

FEDERALLY FUNDED RESEARCH AND DEVELOPMENT CENTER (FFRDC) MODEL

The FFRDC model is based on MIT Lincoln Laboratory, an entity that provides a predetermined level of engineering services through annual line-item funding in the federal budget.

GOVERNMENT-OWNED/CONTRACTOR-OPERATED (GOCO) MODEL

The Government-Owned/Contractor-Operated (GOCO) model is based on such Department of Energy (DoE) GOCOs as Sandia and Los Alamos, where the facility is owned by the U.S. government but operated by a commercial firm under a contract between the firm and the government.

GOVERNMENT AS A SUBCONTRACTOR (GOV SUB) MODEL

The GOV SUB model is a system in which the government laboratories compete with private industry to perform work on government systems. If the government laboratory is selected to perform the work, it becomes an associated contractor or subcontractor to the prime contractor in charge of the government program.

INTERNATIONAL LABORATORY MODEL

In an international laboratory model, scientists from many nations would work together to perform research of mutual interest. Such an international model would inherently offer a high level of leveraging off international organizations because scientists from different nations would work side by side on the same research in an effort to achieve like goals.

JOINT SERVICE LABORATORY (JSL) MODEL

The JSL model is based on the Armed Forces Radiobiology Research Institute, where the Army, Navy, and Air Force combine resources and jointly perform basic and applied research on technologies of interest to all three services.

OUTSOURCE LABORATORY MODEL

The Outsource Laboratory model is based on the Defense Advanced Research Projects Agency (DARPA), which outsources all its research to contractors and academia and uses highly experienced scientists who are government employees (usually term employees) to provide oversight.

PRIVATE LABORATORY (PL) MODEL

In a PL model, the government has no ownership stake and no control over how the facility/business is operated. Ownership of a PL can range from publicly traded stockholder ownership to complete private-party ownership. Government (Army) research can be conducted through charter statements and/or contractual agreements between the government and the PL.

RESERVE SCIENTISTS AND ENGINEERS (S&E) CORPS MODEL

The Reserve S&E Corps model is a system in which scientists and engineers are registered in reserve corps similar to military reserve corps. When the Army requires expertise in a particular area, members of the reserve corps are called upon to perform the services for the government. The S&Es are paid for the services they provide and may receive a fee for being a member of the reserve corps.

INCUBATOR MODEL

A technology incubator laboratory is a lab where an organization, such as the Army, provides basic support services or infrastructure to help fledgling firms develop marketable products. The Army is investing in a firm that it believes is developing a concept that can result in a product or service that the Army can use. The Army can structure incubator agreements for monetary gains and/or priorities in gaining use of the resulting product/service.

VENTURE CAPITAL MODEL

The venture capital model requires the Army to invest in a concept that is not yet fully developed. The money is used to develop the concept into a successful product or service. Such ventures are often clad in some degree of secrecy, and collaborative efforts tend to be limited without an invested monetary interest. When the successful product or service is produced, the investors can receive monetary gains. The Army can receive additional benefits, such as being able to field a system earlier.

VIRTUAL LABORATORY MODEL

A virtual laboratory is a conceptual model, in which S&Es could be located anywhere. The S&Es perform research at local laboratories, are linked via computer and other telecommunication devices, and use these tools to communicate plans, strategies, findings, results, and conclusions, thus enabling them to work together on the same research efforts.

Appendix B

DESCRIPTION OF DEPOT MODELS CONSIDERED

In this appendix, we describe in more detail the depot models used in the analysis described in Chapter Four.

BASELINE PLUS

The Baseline Plus model refers to a restructured and improved depot system operating under full organic control by the Army. It is assumed to be redesigned to seek optimal efficiency and capacity operating within the constraints imposed by the current system. The Baseline Plus model extends the reconstructing commenced by U.S. Army Materiel Command (AMC). These improvements are significant and are key elements of the Baseline Plus model.

Under the new AMC plan signed out by the Assistant Secretary of the Army for Acquisition, Logistics, and Technology on July 29, 1999, the four-star Commander of AMC is to become the Single Manager for Army depot maintenance to "ensure that the organic depot program achieves optimal efficiency and capacity utilization to reduce the depot operating costs." Operating through the office of the AMC Deputy Chief of Staff for Logistics, Headquarters (HQ) AMC will henceforth set policy and control financial execution of the depot system. As part of the change, the five depot facilities will report to HQ AMC through AMC's Major Subordinate Commands (MSCs) rather than continuing to operate under the AMC Industrial Operations Command (IOC) (which will now focus exclusively on the Army's ordnance activities). The new MSC plan also envisions a greater role for Program Managers (PMs) and Program Executive

Officers (PEOs) at the MSCs in controlling the depot maintenance program.

FEDERAL GOVERNMENT CORPORATION (FGC) MODEL

The FGC model is discussed in the main body of the report. The key feature of an FGC is flexibility. FGCs are granted flexibility in regard to otherwise encumbering regulations and in regard to Civil Service rules and federal acquisition and disposal requirements. They are granted freedom from the political forces driving congressional actions. They are allowed to focus on a single product or service and on a limited customer base or constituency by being insulated from the demands of a multimission agency. Finally, FGCs are allowed financial freedoms unavailable to federal agencies. In particular, they can borrow money from commercial sources, they can issue debt in the form of bonds, they can be exempt from local, state, and federal taxes, and they can benefit from "off-the-balance-sheet" status, multiyear federal funding, and exemption from deficit reduction spending caps.

GOVERNMENT-OWNED/CONTRACTOR-OPERATED (GOCO) MODEL

GOCOs are a form of privatization, in which the facilities and equipment are owned by the government but operated by contractors with mostly contractor staffing. They are implemented through standard contracting methods and are entirely supported by the government. Their major advantage is their promise of efficient commercial operation of facilities for which the government is the sole customer, but concerns with oversight and control can lead to micromanagement and declines in efficiency.

GOCOs are a very common organizational structure in the government. For example, all the Department of Energy (DoE) laboratories are GOCOs. These research GOCOs are operated by universities or by university consortia. The contracts involved are generally five-year cost-plus fixed-fee (CPFF). Usually, the government and the contractor have a long-standing relationship built on mutual cooperation and trust.

PRIME VENDOR MODEL

At its most basic, Prime Vendor Support (PVS) is a concept whereby all the wholesale logistics functions for a system are performed under a system contract by a "Prime Vendor." Only operator and unit levels of maintenance are performed by military personnel or other government employees. All other maintenance of the system, repair parts resupply and stockage, reengineering, recapitalization, and modernization is performed by the contractor.

The Prime Vendor concept can be better understood by looking at the proposal for the Apache weapon system. The proposal involves giving a contractor, Team Apache Systems (TAS) (a limited liability company formed by Boeing, Lockheed Martin, and General Electric), total responsibility for wholesale-level logistics support.

By its nature, the Prime Vendor approach is not as much about the disposition of the existing depots as it is about changing the nature of how weapon systems are supported. In that sense, the concept departs from the "ownership perspective." Indeed, part of the reason the Apache proposal ran into problems is because it did not devote enough attention to the question of the effect it would have on the remaining Army wholesale logistics system. To be sure, the Apache proposal does contain provisions for sending some workload to the Corpus Christi Army depot, so it does not completely ignore the existing depot system; nevertheless, the concept is much more about finding new ways to support weapon systems than it is about reforming the way the current physical depot system operates.

PRIVATIZATION IN PLACE MODEL

Based on the Air Force's experience with its San Antonio and Sacramento depots, the term "privatization in place" refers to a private contractor (working with the local municipal authority) assuming ownership of a depot at its current location. Government employees may or may not become employees of the contractor, depending on the arrangements.

A major aspect of the controversy surrounding the San Antonio and Sacramento depots was that by announcing they would be "privatized in place" after the Base Realignment and Closure (BRAC) decisions had been made, the White House, in effect, was denying the three remaining depots in the Air Force the opportunity to add the transferring workloads to their existing work, which denied them the "protection" the additional workloads would bring against their being selected in future BRAC rounds for closure.

All this is germane to the Army situation because the Army depot system, just like the Air Force depot system before the two closures, has substantial excess capacity. The key decision is what to do with the excess capacity: maintain, transform, or eliminate. Privatization in place maintains the excess capacity, while hoping for something that can make use of it.

PUBLIC-PRIVATE PARTNERSHIP (PPP) MODEL

PPPs, which are discussed more in Chapter Two, are a perturbation on the existing AMC restructuring effort, which involves the use of government-owned property, plant, or equipment or the use of government employees to produce goods or services for the private or public sector. In all cases, the private-sector partner also contributes property, plant, equipment, or personnel to achieve the end goal of the partnership. The enabling legislation for these partnering arrangements (10 U.S.C. 4543, 10 U.S.C. 2208(j), 10 U.S.C. 2471, and 10 U.S.C. 2667) states that there can be no other commercial source that could reasonably provide the product in the required time frame and of the required quality and quantity. The 1997 IOC guidance for determining commercial availability required that cost not be a consideration in this determination.

The Army depots have led the way in DoD for exploring this alternative. In particular, Anniston, Red River, and Tobyhanna have ongoing partnerships with private industry in a variety of areas. In each of these cases, the Army has awarded a contract to a private-sector company. The contractor decides that subcontracting with the depot for a portion of the required scope of work would be efficient and statutorily allowed. A contract is signed between the company and the depot, with proceeds paid in advance, which the depot uses to reimburse the working capital fund (WCF). PPPs are one way of "giving back" some work to the organic depot system that is contracted out. Although the opportunity exists in PPPs for the depots to provide goods and services for the private sector, not the government sector, there are very few occurrences of this situation.

The Army also uses work-sharing arrangements with private-sector companies that do not require a contract—only an agreement at the PEO or PM level that the workload will be split. Funding is allocated directly for the depot work and separately for the private-sector work. There are no contracts between the depots and private-sector companies, although memorandums of understanding are exchanged. We exclude these arrangements from our consideration of PPPs because they are an agreement that is included in the restructured AMC depot option discussed above.

BIBLIOGRAPHY

ARCH Development Corporation, *http://www-arch.uchicago.edu.*

ARCH Venture Partners, *http://www.archventure.com/index.html.*

Army Directorate of Public Works, *Annual Summary of Operations, Fiscal Year 1997*, Vol. 1, pp. 2–13.

ArmyLINK News, "Army to Allow Firms to Ship Explosives Through Post," Army News Service, April 13, 2000, *www.dtic.mil/armylink/news/Apr2000/a20000413mtmcexplosives.html.*

Assistant Secretary of the Army, Financial Management and Comptroller, *The Army Budget: FY01 President's Budget,* Washington, D.C.: Headquarters, Department of the Army, February 2000.

Bourgoine, COL Daniel, Deputy Chief of Staff for Doctrine, "Army After Next Insights," briefing slides, United States Army Training and Doctrine Command (TRADOC), *http://www.tradoc.army.mil/dcsdoc/fbdaan/aanframe.htm,* January 1999.

Brown, Edward A., *Reinventing Government Research and Development: A Status Report on Management Initiatives and Reinvention Efforts at the Army Research Laboratory*, Adelphi, MD: Army Research Laboratory, ARL-SR-57, 1998.

Burger, Kim, "Science Board: Technology Readiness Favorable for FCS by 2010," *Inside the Army*, Vol. 12, No. 43, October 30, 2000.

Chang, Ike, Steven Galing, Carolyn Wong, Howell Yee, Elliot Axelband, Mark Onesi, and Kenneth Horn, *Use of Public-Private Part-*

nerships to Meet Future Army Needs, Santa Monica, CA: RAND, MR-997-A, 1999.

Chen, Kathy, "Pentagon Finds Fewer Firms Want to Do Military R&D," *The Wall Street Journal,* October 22, 1999, p. A20.

"Civil Reserve Air Fleet," Defense Daily Network, Phillips Publishing International, Inc., *http://www.defensedaily.com/progprof/usaf/ Civil_Reserve_Air_Fleet.html,* May 1999.

Collender, Stanley E., *The Guide to the Federal Budget Fiscal 1998,* Lanham, Maryland: Rowman & Littlefield, 1997.

Cox, Mathew, "LAND WARRIOR: It's on Its Way. It's Here to Stay. And It Could Forever Change the Infantry," *Army Times,* March 6, 2000, p. 18.

Crawford, Mark, "NRC Identifies Weaknesses in Army Research Laboratory," *Defense Week,* February 23, 1998, p. 15.

Defense Science Board (DSB), *Report of the Defense Science Board Task Force on Defense Science and Technology Base for the 21st Century,* Washington, D.C.: Office of the Under Secretary of Defense for Acquisition and Technology, June 1998.

Directorate for Information Operations and Reports, U.S. Department of Defense, *The Defense Contract Action Data Files, http://web1.whs.osd.mil./peidhome/guide/procoper.htm.*

"Draft Mission Need Statement for the Future Combat System," *Inside the Army,* Vol. 12, No. 2, January 17, 2000.

Dupont, Daniel G., "GAO: Land Warrior Problems May Render Army System 'Ineffective,'" *Inside the Army,* Vol. 11, No. 51, December 27, 1999.

Electronic Data Systems Corporation, "Hoover's Company Profile, 2001," *http://www.hoovers.com.*

European Commission, "Green Paper on Innovation," December 1995, downloadable at *http://www.europa.eu.int/en/record/green/ gp9512/ind_inn.htm.*

European Private Equity and Venture Capital Association, "The Economic Impact of Venture Capital in Europe," 1999, *http://www.evca.com*.

Federal Property and Administrative Services Act of 1949, 41 U.S.C. 251–260.

Froomkin, A. Michael, "Reinventing the Government Corporation," *University of Illinois Law Review*, Vol. 543 (1995) and references therein.

Gladstone, David, *Venture Capital Handbook*, Englewood Cliffs, NJ: Prentice Hall, 1988.

Gompers, Paul A., and Josh Lerner, *The Determinants of Corporate Venture Capital Success: Organizational Structure, Incentives, and Complementarities*, Cambridge, MA: National Bureau of Economic Research, 1998.

Gompers, Paul A., and Josh Lerner, *What Drives Venture Capital Fundraising*, Cambridge, MA: National Bureau of Economic Research, Working Paper 6906, January 1999.

Gonczy, Stephen, "Information Paper on Federal Government Corporations," private communication, September 1998.

Government Corporation Control Act (GCCA), 31 U.S.C. 9101–10 (1988 and Supp. Vol. 1993).

Held, Bruce, and Ike Chang, *Using Venture Capital to Improve Army Research and Development*, Santa Monica, CA: RAND, IP-199, 2000.

Held, Bruce, personal notes, "Perspectives on Other Transactions Conference," 23–24 June 1999, Fort Belvoir, VA, conference sponsored by Defense Contract Management District West, Commander, DCMC Seattle, *http://www.dcmc.hq.dla.mil/ref_info/round/index.htm*.

Horn, Kenneth, Elliot P. Axelband, Ike Yi Chang, Paul S. Steinberg, Carolyn Wong, and Howell Yee, *Performing Collaborative Research with Nontraditional Military Suppliers*, Santa Monica, CA: RAND, MR-830-A, 1997.

Horn, Kenneth, Carolyn Wong, Bruce Held, Elliot Axelband, Paul Steinberg, and Sydne Newberry, *Smart Management of R&D in the 21st Century: Strengthening the Army's Science and Technology Capabilities,* Santa Monica, CA: RAND, IP-210, 2001.

Hynes, Michael, Christopher Hanks, and Ike Chang, "Organizational Alternatives for the Army Depot System," Santa Monica, CA: unpublished RAND research, May 2000.

Hynes, Michael, Sheila Kirby, and Jennifer Sloan, *A Casebook of Alternative Governance Structures and Organizational Forms,* Santa Monica, CA: RAND, MR-1103-OSD, 2000.

Kortum, Samuel, and Josh Lerner, "Does Venture Capital Spur Innovation," Cambridge, MA: National Bureau of Economic Research, Working Paper 6846, December 1998, *http://www.nber.org/papers/w6846.*

Larson, Charles F., *Basic Research for Industry,* Washington, D.C.: Industrial Research Institute, 1999.

Lavine, COL Mike, Chief, Analysis Division, Office of the Assistant Secretary of the Army for Acquisition, Logistics, and Technology, briefing to RAND personnel, May 6, 1999.

Leazes, Francis J., Jr., *Accountability and the Business State,* New York: Prager, 1987.

Lewis, Lawrence, Jr., *A History of the Bank of North America,* Philadelphia: J.B. Lippincott and Co., 1882.

Lilienthal, David E., and Robert H. Marquis, "The Conduct of Business Enterprises by the Federal Government," *Harvard Law Review,* Vol. 54 (1941), p. 545.

Louie, Gilman, President and CEO of In-Q-Tel, interview with the authors, San Mateo, California, December 28, 1999.

Markoff, John, "High-Tech Advances Push C.I.A. into New Company," *The New York Times,* September 29, 1999.

Massachusetts Technology Development Corporation, "MTDC and Its Role in Venture Capital," *http://www.mtdc.com/role.html,* 1999.

Moe, Ronald C., *Administering Public Functions at the Margin of Government: The Case of Federal Corporations*, Washington, D.C.: Congressional Research Service, Library of Congress, 1983.

National Academy of Public Administration (NAPA), *Report on Government Corporations*, Washington, D.C., 1981.

National Research Council (NRC), *The Army Research Laboratory— Alternative Organizational and Management Options*, Committee on Alternate Futures for the Army Research Laboratory, Board of Army Science and Technology, Commission on Engineering and Technical Systems, Washington, D.C.: National Research Council, 1994.

National Research Council (NRC), *1997 Assessment of the Army Research Laboratory*, Army Research Laboratory Technical Assessment Board, Commission on Physical Sciences, Mathematics, and Applications, Washington, D.C.: National Academy Press, 1998.

National Science Foundation, *National Science Board, Science & Engineering Indicators*, Arlington, VA, NSB 98-1, 1998.

National Science Foundation, Division of Science Resources Statistics (SRS), *National Patterns of R&D Resources: 1999 Data Update*, http://www.nsf.gov/sbe/srs/nsf00306/start.htm, 1999.

Office of the Inspector General, *Costs Charged to Other Transactions*, Washington, D.C.: U.S. Department of Defense, Report No. D-2000-065, December 1999.

Office of Technology Transfer, Deputy Under Secretary of Defense (Science and Technology), *Technology Transfer Achievements*, TechTransit Web site, December 1999, http://www.dtic.mil/techtransit/accomp/pla_accomp.html.

Office of the Under Secretary of Defense (Comptroller), *National Defense Budget Estimates for FY 2000*, Washington, D.C.: U.S. Department of Defense, March 1999.

Office of the Undersecretary of Defense (Comptroller), *National Defense Budget Estimates for FY 2001*, Washington, D.C.: U.S. Department of Defense, March 2000.

Open Enterprise, "Price-Waterhouse-Coopers LLP Study Validates the ARMS Program as a Viable Economic Development Tool," *http://www.openterprise.com/newsletter/June99/page1.htm#art1*, 1999.

Owen, David, et al., *Applicability of Alternative Organizational Models to Army Laboratories: A Preliminary E-Delphi Analysis*, Santa Monica, CA: RAND, DB-347-A, 2001 (For Army Use Only).

PriceWaterhouseCoopers, *Eighth Annual Economic Impact of Venture Capital Study*, November 1998.

Rohde, Dr. Robert S., Office of the Assistant Secretary of the Army (Acquisition, Logistics and Technology), *Army COSSI*, briefing slides, n.d.

Sargari, Susan B., *Venture Capital Lessons from the Developed World for the Developing Markets*, Washington, D.C.: International Finance Corporation, The World Bank, Discussion Paper Number 13, 1992.

Schrage, Michael, *How the World's Best Companies Simulate to Innovate*, Cambridge, MA: Harvard Business School Press, 1999.

Shambach, Stephen A., *BRAC Update: Reshaping Today for Tomorrow's Army*, Base Realignment and Closure Office, Washington, D.C.: Headquarters, Department of the Army, December 1999.

Taptich, Brian E., "The New Startup: How the Glut of Venture Capital and the Ripening of the Tech Industry Have Made Entrepreneurs the Bosses of Technology and Changed the Business of Innovation," *Red Herring Magazine*, October 1998.

Truman, Harry S., "Annual Budget Message to Congress: Fiscal Year 1998," January 10, 1947, Public Papers of the Presidents, p. 96.

U.S. Army Research Office, Robert Morris Acquisition Center—RTP Division, "Funding Opportunities," *http://www.aro.army.mil/research/index.htm*.

U.S. Department of Defense, Dual Use Science and Technology Program, "Dual Use Science and Technology Projects from Fiscal

Year 1999," Washington, D.C., *http://www.dtic.mil/dust/cgr/army00cgr.htm#army18*.

U.S. Department of Defense, Dual Use Science and Technology Program, *Guidelines for Dual Use Science and Technology Program, Fiscal Year 2001*, Washington, D.C., September 1999, *http://www.dtic.mil/dust/PDF/01guid.pdf*.

U.S. Department of Defense, *Report on Other Transaction Awards for Prototype Projects; Pursuant to 10 U.S.C 2371, Section 845 (As Amended)*, Washington, D.C., February 26, 1999.

U.S. General Accounting Office, *Congress Should Consider Revising the Government Corporations Act*, Washington, D.C., April 1983.

U.S. General Accounting Office, *Government Corporations: Profiles of Existing Government Corporations*, Washington, D.C., GAO/GGD-96-14, December 1995.

U.S. General Accounting Office, *Army Industrial Facilities: Workforce Requirements and Related Issues Affecting Depots and Arsenals*, Washington, D.C., GAO/NSIAD-99-31, November 1998a.

U.S. General Accounting Office, *Best Practices: Elements Critical to Successfully Reducing Unneeded RDT&E Infrastructure*, Washington, D.C., GAO/NSIAD/RCED-98-23, 1998b.

U.S. General Accounting Office, *Evaluation of Small Business Innovation Research Can Be Strengthened*, Washington, D.C., GAO/RCED-99-114, June 1999a.

U.S. General Accounting Office, *Military Infrastructure: Real Property Needs Improvement*, Washington, D.C., GAO/NSIAD-99-100, September 1999b.

U.S. General Accounting Office, *Public-Private Partnerships: Key Elements of Federal Building and Facilities Partnerships*, Washington, D.C., GAO/GGD-99-23, 1999c.

U.S. General Accounting Office, *Military Housing: Continued Concerns in Implementing the Privatization Initiative*, Washington, D.C., GAO/NSIAD-00-71, March 2000.

U.S. Patent and Trademark Office, *TAF Special Report: All Patents, All Types, January 1977—December 1998*, Washington, D.C.: Office for Patent and Trademark Information/TAF Program, March 1999.

U.S. Patent and Trademark Office, *TAF Profile Report: U.S. (Federal) Government Patenting 1/1977–12/1998*, Washington, D.C.: Office for Patent and Trademark Information, March 1999.

VentureOne Corporation, "1998 Investment Highlights, Venture One Quarterly Statistics," *http://www.ventureone.com/research/venturedata/stats/q498news.html*.

VentureOne Corporation, "Industry Data," 1999, *http://www.ventureone.com/research/venturedata/index.htm*.

Vivar, Jonathan, and James Reay, *Applying the Government Corporation and Other Organizational Concepts to DoD's Defense Working Capital Fund*, Washington, D.C.: Logistics Management Institute, DR901T1, May 1999.

Washington, W. N., "Depot Utilization and Commercialization," *Acquisition Review Quarterly*, Summer 1999.

Winograd, Erin Q., "AMC Chief Advocates Strong S&T Investment as Key to Future Readiness," *Inside the Army*, Vol. II, No. 41, October 18, 1999.

Wolfe, Raymond M., *1997 U.S. Industrial R&D Performers*, Arlington, VA: National Science Foundation, August 1999, *http://www.nsf.gov/sbe/srs/nsf99355/pdf/nsf99355.pdf*.

Wong, Carolyn, *An Analysis of Collaborative Research Opportunities for the Army*, Santa Monica, CA: RAND, MR-675-A, 1998.